THE FUTURE OF HUMANITY IS NOW

Where Are We Going in The Future (As A Species)?

Melvin Dodson III, M.D., PhD

Copyright © 2024 by Melvin Dodson.

All Rights Reserved. This book or any portion thereof may not be reproduced or used in any manner whatsoever without the express written permission of the author except for the use of brief quotations in a book or book review that are appropriately referenced.

The information herein represents the view of the author as of the date of publication. This book is presented for informational purposes only. Due to the rate at which conditions change, the author reserves the right to alter and update the book based on new conditions. While every attempt has been made to verify the information in this book, neither the author not his affiliates/partners assume any responsibility for errors, inaccuracies, or omissions.

Dedication

THE RESURRECTION IS COMING!

This book is dedicated to mankind, with the fervent hope and sincere desire that mankind will, in the not-so-distant future, find real wisdom and knowledge, tranquility with peace on Earth, and true enlightenment with advanced intelligence. Only mankind can bestow upon us these virtues from an abstract metaphysical "Heaven," which has in the past always been just beyond mankind's reach and reality.

Mankind must come to the realization that seeking happiness and immortality is a never-ending journey that can only be realized after we have climbed Mount Olympus and discovered that **the Gods that we seek, and the Gods that we have lusted for since our trials and tribulations as cavemen, are none other than the future development of ourselves. We are the now and future Gods!**

Kneel and give thanks, you stand on sacred ground…mother Earth, the giver of life. Scientists tell us that there is a very high probability that there are other intelligent life forms in this Universe. Despite the "high possibility" none has stopped by to say "Hello" (or at least none that we know of for sure).

If you want to Worship the one and only true God…at the moment, that appears to be Mankind. Mankind and our progeny are the only Gods that will ever walk the face of this Earth. In addition to creating all the Gods of our forefathers, mankind even invented (created) the basic concept of Gods. **We (mankind) are the only God(s).**

In addition, mankind is at the point of a paradigm shift in evolution. After almost 4 billion years of evolution driven by random mutations and natural selection, **mankind is about to take over our own evolution, and turn it from a pointless, blind, pitiless and indifferent process into an intelligently directed process for the first time...ever.** This will result in the most extreme developments in Human evolution that has ever occurred! Incredibly, evolution for mankind is at the beginning of a revolution that will incredibly affect mankind's future!

"Alas poor Yorick, I knew him" well; he is but flesh and bone, yet he will arise as a God! Keep the faith. The Resurrection is coming.

Blessed be the Gods!

TABLE OF CONTENTS

Dedication: The Resurrection Is Coming! iii
Where Are We Going In The Future (As A Species)? vi
Preface viii
Chapter 1: Our Future 1
Chapter 2: Armageddon: The Future Extinction Of Homo Sapiens 24
Chapter 3: Intelligent Directed Evolution By Humans (Ide_h) 60
Chapter 4: Methuselah 135
Chapter 5: The Wise Monkey 203
Chapter 6: The Dumb Machine 299
Chapter 7: On Our Way, Homo Intelligenticus 343
Chapter 8: The Final Frontier, The Future Of God 349
About The Author 367

WHERE ARE WE GOING IN THE FUTURE (AS A SPECIES)?

Chapter 1: Our Future
- We have come a long way Baby, but there is still a winding and rocky road ahead.
- We must take that road with vigor, because the only other road leads to our grave and extinction!

Chapter 2: Armageddon: The future extinction of Homo Sapiens
- We live in a tough neighborhood.
- A history of Five mass extinctions, and perhaps we are living in a sixth mass extinction of our own making.

Chapter 3: Intelligent Directed Evolution by Humans (IDE_H)
- Man, the maker of mankind.
- Intelligence, do not leave home without it.

Chapter 4: Methuselah
- Mankind must choose: The Heaven of non-existence in a mythological cloud or Immortality on Earth. The Earth is our Heaven. The Earth is our Hell. Upon this rock, ye shall build your house.

Chapter 5: The Wise Monkey
- Intelligence is the foundation upon which mankind must build his house. Intelligence is the essence of Gods.

Chapter 6: The Dumb Machine
- Computers and Artificial Intelligence; we built them, and they are us!

Chapter 7: On Our Way, Homo Intelligenticus
- Are there space men? Yes, we are the Space Men!

Chapter 8: The Final Frontier, The Future of God
- Are there Gods? Yes, we humans are the now and future Gods!
- Mankind is the God Creator! Mankind even created and defined the concept of Gods!

PREFACE

The Triquandry

There are many basic and profound questions that are fundamental to the understanding and appreciation of Human existence. We are going to look at and examine question of mankind's future.

The Future Is Now!
Where are we "going" (read: "evolving as a species") in the relative near future (the next several hundred years)?

These questions sum to the "story" of the incredible and almost unbelievable future evolution of mankind and even potentially, the incredible future development of mankind's intelligence. The possibly (hopefully) of even change classic Darwinian evolution from a process of random mutations to an **intelligent directed evolutionary process, directed by Human intelligence.**

Our forefathers projected many concepts of creation. The most popular, at least in the West, comes from the Bible. This has been characterized as **"the greatest story ever told."** It has been mankind's perception and consideration for our origin and the origin of everything in the Universe for several thousand years. Unfortunately, it has persisted as a compassionately held belief system, derived from religions, but contrary to known facts developed from science.

THE OTHER Greatest Story Ever Told is an alternative concept of creation that has developed from science. ***THE OTHER...*** is an epic in the true sense of the term, with the Universe, stars, life, mankind, intelligence, with past, present and future as some of the evolving hero of this epic.

Life appears to have begun as an emergent phenomenon originating from the combinatorics of atoms and molecules into a multitude of self-replicating, energy-consuming entities that have continued to evolve by random changes into the myriad forms of life on Earth, limited only by the environmental constraints, and the physical laws of the Universe. Life now occupies every nook and cranny of possible existence on the planet Earth. This has been a journey without a plan, but guided by the invisible hand of what is physically possible and restricted only by the laws of physics.

The only criterion of success is continued existence, reproduction, and survival. From such simplicity and focus arose multiple other emergent phenomena, where the whole is more than the sum of the parts. However, the most incredible emergent phenomena in addition to life itself has been the development of advanced intelligence, which has nurtured the understanding of itself and everything around itself. This second "creation" event rivals in magnificence the creation of life itself.

However, as time ticks to eternity, the future will reveal other emergent phenomena with new heroes that can be visualized as **"evolving new Gods."** You do not believe in Gods? Look in a mirror and I will show you a God or at least an evolving new God! **Yes, pitiful, puny, and even at times perhaps...pathetic, but an evolving new God no less.** Most readers will be non-believers in mankind as an evolving now or perhaps even a future God. However, it is all in the definition and understanding of: "what is a God?" It is also all in the understanding of the evolutionary process itself

Mankind is the God creator, we have created thousands of Gods: Egyptian Gods, Mesopotamian Gods, Greek Gods, Roman Gods. Some have been simply stone carvings on a wall. Some of the Gods have been Chimaeras (Frankenstein-like Gods, with a piece from here and a piece from there, such as Thoth [the Egyptian

God of writing with an Ibis head and a Human body]), however many have had very Human forms, and at least one God is totally invisible with (apparently) no body at all (the God of Abraham). This was particularly convenient for the descendants of Abraham, since they kept the creator God in a box or on top of the box or simple near the box. The box was easily carried around. The Box contained the 10 commandments and was easily carried around (The Arch of The Covenant!). Although it is not clear exactly where God was in relation to the box. One possible suggestion is that the God of Abraham is everywhere, and cannot be localized, and the Arch of the Covenant simply contains the ten commandments, and is simply a "symbol of the God of Abraham". However, the JEWS kept the "ARC" in the "Holy of Holies" in the temple in Jerusalem and treated it as if it had some special relationship to the God of Abraham!

The Jews were nomads, and therefore such a God was both locally pragmatic and efficient.

In addition, mankind is even the inventor or creator of the concept of a God itself. No other lifeform besides Humans that we are aware of has created a God. In the absence of mankind, the concept of God does not exist. No other creature on Earth worships Gods or even understands the concept of a God. It is clear that **Mankind is the God creator**, but there is one God left for mankind to create or at least perfect... mankind himself. Mankind is well on our way to achieving this goal using new knowledge and new technical skills.

Those that read and understand this book will have at least a reasonable basic understanding of the origin of the Universe, the emergence and struggle of life and pursuit of existence, the emergence of Humans (at least in outline form), and equally as profound, the emergence of intelligence that has created the search for the meaning and purpose of life.

However, the journey continues! There is an absolute necessity for mankind to take over our own evolution and secure a future that will rival the mythology of mankind's many Gods. The necessity for mankind to direct our own future evolution in an intelligent manner rather than evolution by blind random mutations is critical for our future, and even our survival. This can be readily appreciated when we note that more than 99% of all the varied species of creatures that ever existed on Earth by some estimates are now extinct. In addition, the future of every Human on Earth is death..."thank you Mother Nature." But what else would you expect from a random mutational process that has no knowledge, no consideration for the individual involved. Evolution is simply a physical process. The simple fact is that Humans have evolved advanced intelligence, and has come to understand a great deal about the Universe in which we live, and also about ourselves. However, we have only minimal control of our environment, and have only a brief existence, and then we perish. When we are "gone," we are gone forever. There is no Heaven, and there is no afterlife. Our forefathers have created a fairy tale future for mankind that does not exist. This book is about understanding the reality of our existence, but also about the incredible future possibilities.

Simply stated, this book has the power to change your understanding of mankind's place in the Universe, and possibility even your life or at least your attitudes and perspectives of mankind's future...don't leave life without it.

"I love this book."

"What an egotistical statement for an author to make."

"Not really…let me explain."

I love this book because writing it has been a wonderful adventure of the "mind-kind" for me personally. It has been a nirvana of understanding for me personally. You really can never go any further in this world than where you go in your own mind. This book has taken me from the farthest reaches of outer space to the atoms and molecules of life. This has been a personal journey for me, and I hope that it will be a personal journey for you too.

We ask three simple, yet profound questions.

However, the answers will take us on a journey from the cosmos to the infinitesimal, from the beginning to the present, from the ridiculous to the sublime and even extent into the unknown future. It will be one Hell of a journey, yet you will never leave the three-pound Universe that lies between your ears. In addition to an adventure, it's also educational and gives a minimally acceptable understanding of some physics, astronomy, cosmology, biology, evolution, paleontology, psychology, philosophy and computer science, and even some medicine (in the area of aging). Consider this a college course: "Introduction to Mankind 101." Do not fear education and understanding, because ultimately it is your road map for our Universe and life itself.

You are on a journey through life…it began at conception. Life was consecrated on the day you were born, and death is likewise assured. You can have a road map that takes you in circles in life and has many blind-ends. You can seek the way of the wise traveler, who knows the pitfalls that lie in the road ahead…and that also lie in your minds-eye as you see your way through life. **Take my hand, and I will walk with you in this journey through life and understanding.**

"Where did we come from?"

The answer is difficult, complicated and follows a circuitous path that meanders through the Universe and disappears into the atom. However, anyone that considers themselves to be a Human should seek an answer and have at least a basic understanding of this fundamental question. The answers are embedded in what makes you (us) a Human. The answers are enlightening, and even a superficial understanding can literally illuminate one's life. The answers will free oneself from a life of slavery to mythology and ignorance.

"It is most holy." (Bible, KJV, Leviticus 6:29)

Yet, incredibly, the many answers that we will review could not have been adequately and accurately answered as little as one hundred years and fifty years ago or so, and prior to 1859 (the publication of Darwin's The Origin of Species), mankind did not even have a clue as to mankind's origins, the origin of life, or even the origins of the Earth or Universe. We did not understand or even vaguely appreciate the incredible and magnificent Universe in which we live. In fact, until recently, we really did not even vaguely understand the composition of the Universe or ourselves, nor did we appreciate that we are a mere speck in an incredibly huge Universe. We are in fact privileged to even be able to realistically examine, study, and consider these questions. The answers have come slowly. Men (women) have paid the ultimate price for the knowledge and understanding in these pages that follow. Open this book and the blood of the martyrs for knowledge and understanding will figuratively flow out of it.

"Verily, Verily, I say unto you…you are blessed among all the men who have walked the face of the Earth, for you will eat of the apple from the tree of knowledge of good and evil and you will be enlightened."

"It is most holy." (Bible, KJV, Leviticus 6:29)

In order to answer the question: "Where did we come from?" we will be taking a journey from the beginning of the past into the future and quite probably to the end of man…at least the version of mankind that we know today. These are all a part of the Holy Grail which this book seeks. This book is a journey from the beginning of the beginning to the now of man until the baton of intelligence is passed on to a more suitable being—a new species…our children…of sorts.

A "tall order"—yes; it needs to be! It is a "tall" story!

Although the title of this book is Think Incredible Thoughts, I have immodestly dubbed this book *THE OTHER Greatest Story Ever Told*…because it is! It is our story. It is the story of our Universe…the story of life on Earth…the story of mankind. It is also why we are here, and it is a road map to where we will be going in the future. However, this is not a fairy tale.

In fact, I do not think anyone on Earth prior to the twentieth century could have truly envisioned or even fully understood this story. They may have seen a small facet or one piece of the puzzle, but not the full story, and especially not its implications. This story is very different from the **"Greatest Story Ever Told"** as depicted in the Bible. That is why it is **THE OTHER...** However, as we proceed, we will periodically stop and compare these two "stories." The two stories are intimately intertwined in the chronicle of the epic and saga of mankind.

We will consider man as a **work-in-progress**, which implies an earlier mankind and a later mankind, which are not the same mankind. We will encounter new possibilities' in this book. These themes are what might be considered the axioms at the base of Human experience. We will see the **Dilemma-of-Man.** We (man of the twentieth and twenty-first century, and to some extent our more recent ancestors of the past 5,000 to 10,000 years or so) are the "sacrificial lambs" of intelligence. We know too much, but not enough…to save ourselves. We now know enough to "figure it out," and to answer profound and basic questions, such as where we came from and how we got here. In fact, the last several hundred years have resulted in a huge tsunami of information rolling over mankind, and an incredible epiphany of understanding about the Universe in which we live. However, we still do not understand enough to alter the reality of our destiny. We are the sacrificial lambs on the altar of intelligence, and we are even smart enough to know it. We wish to survive in a non-survivable world, but we understand that we cannot survive, yet we have a "survival instinct" built into the very nature of our being…that is the **Dilemma-of-Man.** We have an "instinct" or basic emotional drive to survive, but we have evolved to a point of intelligence that we understand survival is not possible…that is our dilemma. We are the only living creature on Earth who understands the inevitability of death and non-existence. Our intelligence has betrayed us. We pray to God! In fact, it is the **Dilemma-of-Man** that is at the base of most religions, including Christianity. We are intelligent. We understand the world (at least to a much higher level than any creature before us), but most important we understand the inevitability of the end of our personal existence.

Men in the past have cleverly developed an answer. There is an afterlife (possibly invented in part by the Egyptians). This afterlife became the Heaven of Christians. We will never cease to exist; we will live on for an eternity. There are, of course, a few caveats required for this scenario, such a believing in certain Gods and acceptable behavior, etc. Unfortunately, we have solved the dilemma created by our advanced intelligence and understanding versus the reality of existence by "dumbing-down" our intelligence and accepting fairy tales as reality. However, who cares? We Humans have now become immortal. Our existence will not end when we die! How absolutely convenient.

"Is there meaning and purpose to Human life?"

This is our second question, and it is even harder to answer, and the answers are certainly even more contentious. Perhaps there is no answer. Perhaps the answer must come from "on-high" as it has in the past. Perhaps the answer is within us. We shall see!

In our journey, we will consider God. Realistically, we have no choice. God is so interwoven into so many of the places we will visit that not to consider God would be a malfeasance of our intellectual journey. We will journey from the abstract metaphysical reflections of God, morals, and sin to the interesting and provocative considerations of the origins of Homo sapiens—ourselves. Does God even exist? We will also be studying this rather profound question in detail, and the answer is clearly and without question…yes, but just as clearly and without question, the God that exists in reality is not even vaguely the God of our forefathers.

We will also look at morals and ethics and even sin. Why, because these subjects are an integral part of the story of man—particularly when we begin to look for meaning and purpose in life. Mankind is the only creature on Earth that literally demands a meaning and purpose for life…but only Human life. Have you ever heard of "a Frog Heaven"?

Man has transcended the mundane and journeyed into the metaphysical and the abstract, yet we still live in the mundane world, and our daily needs are anything but abstract. In addition, we are inseparably tied to the realities of the basic biology of life and the physics of existence.

Mankind is the only creature that literally demands a meaning to life and existence. Yet, we live in what Richard Dawkins described as a Universe with "no purpose, no evil, no good, nothing but blind, pitiless indifference." However, we are a sentient creature that has begun to take existence beyond the mundane. Can we give meaning and purpose to Human life in a pointless and purposeless world? Again, **yes**, but it will not be the meaning and purpose of our forefathers. I beg patience, until we address these issues.

We will look at the **Manifest Destiny of Mankind**. A prophecy for man's prospects in the still unknown—the future. The **Manifest Destiny of Mankind** is a road map to the meaning and purpose of mankind's future existence. **Do not fear; understanding will be our sword and our shield for this fight for the future!**

"Yea, though [we] walk through the valley of the shadow of death, [we] will fear no evil" (Psalm 23:4). **Although the bodies of mankind litter the road of evolution, the Promised Land has always lied ahead in the future.**

As an aside, I confess that I have stolen a writing technique from the early Greek playwrights with a Greek choir off-stage. When a Greek playwright wished to comment on something, a Greek choir (who was off-stage and could not be seen) would sing the refrain. These comments sometimes appear to be coming from "on-high." Sometimes our "Greek choir off-stage" will sound like the consciousness of mankind. This choir will be our Jiminy Cricket of Pinocchio fame, our soul…of sorts. Sometimes our "Greek choir" (our Jiminy Cricket) will emphasize a point. Sometimes they may taunt us. I will use the term: **Greek Choir Off-Stage (GCOS) in bold to introduce them. Sometimes, I will comment back to the choir. My comments will be in quotation marks.**

GCOS: "Shall we speak in parables?"
"It worked for Jesus."

Consider a herd of cattle grazing in a meadow. It is spring, and the weather is pleasant; the grass is plentiful. There are calves near their mothers. There are no predators. The cattle are peaceful. Life is in harmony. The cattle graze: they feed. They reproduce. Life goes on. Their world extends only to the edge of the meadow.

The cattle in this pleasant pasture have no knowledge of any of the concepts reviewed in this book. They have no idea where they came from, where they are going or why there are "here." They have no awareness of atoms or the composition of stars: no concept of philosophy, science, mathematics or metaphysics.

GCOS: "Of course they know none of these matters… they are cattle—only dumb animals."
"They feed. They reproduce. Life goes on."

One day, a truck comes into the pleasant pasture. The cattle are loaded onto the truck and taken to a slaughterhouse. They are fulfilling their purpose in life.

Yet, they never knew: "Why they were here" or what their "purpose" in life really was. In essence, there was no "purpose" for them. Their world extended only to the edge of the grass.

GCOS: "Ignorance is bliss!"
"Perhaps…for some!"

For some life forms, a pleasant meadow is all that exists. It is all that there is or ever will be…for them.

Yet, as noted above, for a few (one species only) … there has been an incredible epiphany of cosmic cognizance. However, this blessing is also a curse. This is particularly true for a creature with some advanced intelligence that is smart enough to understand some of the problems, but not quite smart enough to understand any of the solutions. That creature would be an intellectual sacrificial lamb. The current Homo sapiens is that creature! We are the sacrificial lambs on the altar of knowledge and intelligence. We know too much, but we can do nothing but pray to an invisible God for salvation. The Gospel of John calls Jesus the "Lamb of God," who was sacrificed for the sins of mankind. Perhaps all of Humanity is that same "Lamb of God."

GCOS: "Verily, Verily, I say to you: 'Mankind is the Lamb of Intelligence being sacrificed on the altar of reality.'"

Knowledge has grown from the mustard seed into a gigantic tree. Make no mistake—the growth of the tree of knowledge has been watered by the blood of many martyrs. Knowledge has not been cheaply won. You are privileged to live at a time when the tree of knowledge has brought forth fruit. You are privileged to live at a time when mankind is beginning to walk out of the fog of the road of ignorance behind us, and before us… the road is clear.

GCOS: "Mankind…behold thy world."
"Wait…this does not look like the garden of Eden!"

GCOS: "Mankind…this is the only Garden of Eden that you will ever know…this is your Heaven…this is your Hell…this is your Earth!"

"Where are we going in the future (as a species)?"

This question is very interesting, because it is our future. Where we have been is a matter of history. Sometimes that history has had to be dug out of the ground—literally. Sometimes it has come from a scientific laboratory, but it is history all the same. "Where we are going" is still within our purview. We can still beat the ass of present existence on which we ride and drive it to our chosen destiny…maybe! However, the future is intimately embedded in our past and in our reason for existence.

GCOS: "As you sow, so shall you reap." (A common shrewd and astute philosophical insight paraphrased from Galatians 6:7, Bible, KJV)

We are at or near a "critical point" of Human evolution, where man may be on the brink of remaking man "in his own image." We may be at the point of evolving into a new species. This is particularly true in the area of intelligence, which will be the defining character of future man. We will look carefully at the prospect and necessity for mankind to assume control of our own evolution. Taking over our own evolution using genetic engineering is not just convenient; it is vitally necessary both for the survival of mankind as a species, and for any future advancement of mankind. The road to mankind from the past is piled high with the corpses of failed evolutionary experiments (or acts of God if you prefer). We will all age; we will all die within a pitifully short lifespan. This is our "gift" from "Mother Nature" (or God). Taking over our own evolution is the only real salvation for mankind. The Egyptians were possibly the first to invent the afterlife, and they built huge pyramids as "tools" to project them into the afterlife and the world of the Gods. Modern man has recognized the pyramids as "fool's gold" for a journey to the afterlife. In addition, modern Western mankind has built magnificent architectural masterpieces like the Cathedral of Notre Dame...more "fool's gold." Our churches, as magnificent as they are, have worked about as well as the pyramids in providing an afterlife in Heaven. Churches are stone and mortar, and they are not the direction to Heaven...or anything else for that matter.

GCOS: "Wake-up mankind! You are the most intelligent creature on the face of the Earth! If you want a Garden of Eden or a Heaven, you must build it yourself! Forget about children's mythologies and fairy tales. You are the only God on the planet Earth. If you want a heaven, then you must build it! No other "God" or fairy tale is going to give it to you."

We will look realistically at the prospects of extending the lifespan of man for a thousand years or perhaps even indefinitely, and we will consider the realistic prospects of increasing the intelligence of mankind to the point that future mankind…our children, can "eat from the tree of life," and enjoy the fruits of the "knowledge of good and evil" on the only Heaven we will ever know, our "Garden of Eden" …the planet Earth, filled with Gods…ourselves.

Most of the Humans on Earth have no knowledge of many of the concepts discussed in this book.

GCOS: "They feed. They reproduce. Life goes on."

"Are we Humans simply living in a pasture like cattle, waiting to be harvested by the Great Reaper?"

Even among the educated few—they have "heard" about some of the "things" in this book, but the concepts are vague…distant. When I have asked some well-educated friends (generally non-scientists) a simple question such as: "How old is the Earth?"

Most have only a vague idea (I exclude from this comment 12-year-olds who watch the Discovery Channel) …

"Guess…"

"Ten million years."

"I'll give you a clue. I have a dinosaur bone (upper tibia) in my living room that I use as a bookend (and occasionally as a doorstop) that is at least 55 million years older than that."

I am appalled at the number of educated adults who know so little about where they came from, how they got here, why they are even "here," and certainly have no insight into where we (the monkey called man) are going (as a species). The great majority of the world's population considers the past history of mankind with answers that are some variations of a number of popular fairy tales.

This is one of the myriad reasons that I place pen on paper.

Humanity is not just whirling through space and time on a small rock called Earth. There is hope for mankind! Great "things" lie ahead!

However, there are over 7 billion people living in the world today, who are living in a mythical land that bears only scant resemblance to the reality of current Human existence. We ask the question: "Can the most intelligent creature on the planet Earth really be that naive, ignorant and dense?" Some reality is known, but it is generally ignored.

We also must recognize the dichotomy between what is, and what can be.

Another reason I write this tome is: **"I have to!"** This book is literally demanding to be written, and I would fear for my sanity should I not heed my inner voice.

"Will it never end?"

This book urges you to do as Martin Luther King Jr. admonished: have a dream…climb the mountain. See where we have been and look to the horizon…to see where we are going. Do you see the Promised Land?

Now my apology! No one can truly master the many and varied subjects of this book. No one can really write this book—including myself. There is no authority—no man or woman on the face of this Earth that knows or fully understands all the subjects of this book. Perhaps only a fool would even try to write this book.

"I am that fool… I do apologize!"

I do this deed because I believe that this story will always be with us. This story desperately needs to be understood and told to everyone claiming to be a Homo sapiens—not just in this book, but again and again, over and over, better and better, in ever more detail… hopefully by others in the future. Take what you can from this book, steal what you may, but always keep open the "Door of Inquiry" into the quest that we seek here…the story of mankind—

THE OTHER Greatest Story Ever Told.

If this book is not rewritten periodically in the future by others—hopefully by more talented writers than myself, there will be a curse upon mankind…**the curse of ignorance.**

We have taken a bite out of the apple from the tree of the knowledge of good and evil. "Behold, the **man has become one of us, to know good and evil**" (my bold), (Bible, Gen. 3:22).

Now we must eat the whole apple and all the rest of the apples on the tree! Otherwise, we commit the sacrilege of the "final sin" …ignorance for eternity, intelligent life trapped in a quicksand mythology.

To eat from the tree of good and evil is to emerge as a sentient being…to become Human! That is what we have done. Genesis is correct, we have sinned, but sin has made us Human. Without that "bite of an apple," we would be no more than every other living creature, past and present, on the planet Earth. We have sinned! But it is sin that has raised us above the worm.

There is another tree in the Garden of Eden…it is called the tree of life (it is mentioned in Genesis). We have not eaten the fruit from the tree of life…yet. **But we shall eat! The Tree of Life will make us free!** After we eat, we will recognize the real nakedness of our existence. This book predicts that mankind can and will have a glorious understanding in the future that will make us into the Gods that we have created in our mythology. Mythology will become reality.

We have only recently begun to recognize that the greatest threat to mankind is aging and death. We age; we die, because evolution has no use for a wise old monkey. However, when we recognize that we no longer have any use for "Mother Nature's" evolution, then we will become masters of our own destiny…then we will become the "real" Gods that we have always lusted to be, since the beginning of mankind. **We are Gods even now!** Yes…a puny, pathetic, pitiful, inept and inadequate God, but a God, nonetheless. What other creature on Earth understands both the stars and the metaphysics of mathematics? What other creature even vaguely understands the concept of a God? We Humans invented the concept of God, and we are now evolving into Gods ourselves, as we take control of our own evolution.

But the future holds a cornucopia of the fruits of the Gods. Mankind must emerge from our delusions and use our intelligence to build a road to the future. Failure to recognize the dangers of ignorance can be lethal…ask any dinosaur. We live in a tough neighborhood. We need to "think" our way into a new future. What awaits us currently is only our grave and the abyss of nothingness. Nothingness is a much worse fate than HELL!

Liberty, and freedom of our mind and intelligence must reign, but they will not do so until we extract ourselves from the yoke of death, and the tyranny of age and disease, and grasp the full realization of the reality of the fact that that belief and dependence on a God from "on-high" varies in inverse proportion to the intelligence of mankind. How to eat the fruit of the **"Tree of Life"** that still awaits us in the Garden of Eden is the quest of the last section of our journey.

GCOS: "They feed. They reproduce. Life goes on."
As the ancient sages of antiquity have noted: every journey begins with a first step. However, our journey will not be one of many steps, but a journey of a myriad thoughts. **Take my hand…the journey is but a thought ahead. We will think together.** The only risks that lie ahead are the infinity of thoughts and ideas…in your(my) own mind(s). **Failure to take this journey is not an option…it would be a catastrophe for you specifically and for our species in general!**

What Humanity needs is "thought masters" for the future.

What thoughts?

Great thoughts! The incredible thoughts of the Gods! What will save Humanity in the future is YOU (and I) as Thought Masters!

This book is a journey like…life itself.

Melvin Dodson III

I will think through this journey with you, and I will be with you at "the end!"

Your humble servant and guide:
Melvin Dodson III, M.D., PhD

Chapter 1
OUR FUTURE

In reality, the answer to "Where are we going in the future (as a species)?" is wherever we want to go! The future is literally in our hands!

However, mankind's apparent inability to clearly plan for an intellectual and physical future makes the outcome difficult to predict.

Where are we going in the future (as a species)? What is the future of mankind in this Universe? This is the last of our important questions. We are approaching the end of our journey. We see the Holy Grail ahead. It lies in the future.

My father died when I was five years old, and I was raised by my grandfather and grandmother. My grandfather was born in Ohio in 1886, and passed away in 1967, age 81. When I was young, he often told me about all the incredible changes that had occurred in America during his earlier years. It was clearly an impressive time. When he was young, he traveled by horse and wagon. Crossing a river sometimes meant finding a shallow part of the river or a ferry. There were dirt roads; not nearly as many bridges in the Midwest as today (LIKE IN Ohio and Indiana). Only shortly before he was born, Thomas A. Edison had marketed the first commercially successful incandescent lamp (around 1879), seven years before my grandfather was born, but the spread of electric generators and electric lights across the country was still a phenomenon that my grandfather lived through during the late 19th and early 20th century (actually Edison did not "invent" the electric light bulb, however, he improved on it, and he made it a commercially viable enterprise. Edison bought the patent for the electric light from Henry Woodward and Matthew Evans who had patented it in 1875 (but could not bring it to market). It was the age of electricity and electric lighting.

My grandfather told me of walking to school in the snow with no shoes. He recalled days when there was little or nothing to eat. Most people were poor in America. My grandfather was poor, and he never finished the third grade. He had to work. Yet, during all the time I can remember, which were my grandfather's later years, he drove a Cadillac, and we lived in a very nice house, and we had a separate winter home in Florida WHICH HE OWNED. I never was without shoes, and I never went hungry. The roads were paved from one end of the country to the other, even when I was young (I was born in 1940).

What about the invention and marketing of all those incredible things like refrigerators? I can still remember the iceman coming to our house when I was little. He would carry a large block of ice on his shoulder, which was covered by a large piece of leather, using a metal ice tong, and he would put the block of ice in our ice box. However, this soon changed, and there were electric refrigerators, dishwashers, clothes washers, dryers, toasters, electric stoves, and the list transformed our house and our lifestyle when I was still very young. When I was young, I listened to the radio. There was no television. The first time I ever saw a television, it was at a friend's house whose father was a lawyer. I was in the third grade.

I remember many of the things my grandfather told me that he lived through when he was young, and I never thought that I would see the incredible changes in my lifetime like my grandfather had seen. I was wrong. All these changes did not exactly happen overnight, but they did happen in a period of only a hundred years or so; when he was young (before the turn of the century) and later when I was young.

The end of the 19th century and through the beginning of the 20th century was an incredible time. I can still recall our drives to Florida for the winter. Just north of Brunswick, Georgia, there were several wooden bridges, maybe 30 yards in length. There was a single lane for traffic ON THESE BRIDGES. The bridge was too narrow to allow two cars to cross IN OPPOSITE DIRECTIONS at THE SAME TIME time. You had to look and make sure that another car had not started to cross from the other end of the bridge, because only one car would fit on the bridge at a time, and then…just barely. When we crossed the bridge, my grandfather's Cadillac shook the bridge. There were no cement bridges when he was young (before the turn of the century), and even in some places when I was young. You drove directly on the wooden beams of the bridge. They looked like railroad ties, but a bit longer, but no wider or thicker.

The last three or four years before they finally replaced the bridge, at least two of the ties in the bridge were missing, leaving a sizable gap that the car's tires fell into. That shook the car even more than usual when you crossed, and I held on tight. I guess there really wasn't a lot of danger, but as a young child, it really impressed me. Sometimes, I wondered if we were going to make it. Was this a small side road in rural Georgia? No… this was U.S. One. This was the major highway running from New York to Florida in the 1940s (and still is today). But in a short while that was all rebuilt, and there were superhighways. There were bridges over every river, and almost everyone seemed to have a car. It was an economic miracle—a fulfillment of the "American Dream." It was the age of the automobile! Is that what the 20th century will be called—the age of the automobile or the economic miracle in America? Wait, there was more. I remember many more incredible things happening in my lifetime.

I remember Gramps telling me about the first time that he ever heard a radio. I listened to the radio frequently as a child. There was a program called Inner Sanctum. The program opened with this eerie squeaking door. There was nothing scarier than Inner Sanctum…before or since. Although…perhaps it probably helped that I was little, in kindergarten or perhaps first grade.

Gramps and I saw the introduction of television together. The first television in my neighborhood belonged to an attorney's son who lived a couple of blocks away from us in Miami Shores, Florida. All the kids in the neighborhood went to his house to watch the afternoon serial adventure. Television was in black and white, and most of the time all you would see was this test pattern until five o'clock. But soon, we had a television, and so did everyone else, and before too long…it was in color. Even the test patterns were gone in the afternoons. There was television all day. It was the age of television!

Modern medicine was revolutionized during the 20th century—antibiotics, blood transfusions, modern surgery, immunizations were widespread, and probably the most important advancement was a much better understanding of diseases and their causes. There was an endless stream of new and life-saving medical advances. The average lifespan of man expanded by over 30–40 years (depending on the interval being considered). Lifespan more than doubling compared to earlier centuries. My Grandfather had developed pneumonia as a young man—a deadly disease at the time. He was one of the first people in his town to receive treatment with a new sulfa drug. He felt it saved his life, and quite probably it did. My grandmother told me once that if you got pneumonia, you would almost always die. Pneumonia before the advent of antibiotics was a killer. The 20th century was the age of modern medical miracles and the age of antibiotics! People lived through pneumonia! They also began to survive appendicitis with surgical intervention. People survived many other infections as well. It was the miracle of modern medicine. Life expectancy increased dramatically as noted above.

I remember in the 1950s it was the atomic age. Unfortunately, the Second World War had ended with a bang, a nuclear bang. We now controlled the atom or at least enough to make a big noise. There was going to be unlimited power from nuclear energy. Unfortunately, as it turned out, in retrospect, it appeared more like "the atom controlled us," rather than "we controlled the atom". The Cold War and the nuclear arms race with Russia began almost immediately after the introduction of the atomic bomb. Was there a time of peace after the Second World War? You'll have to forgive me, I was small…I must have missed it.

Perhaps my grandfather remembered that time of peace. I do remember drills in grammar school where we all hid under our desks, and we put our arms over our head. You were not supposed to look at the light. I didn't really want to see it anyway. But I do remember very clearly that I was living in the Atomic Age, the 20th century. But the only thing that the Atomic Age was doing for me at the time was the emergency drills in school where we were all hiding under our desks. It did not seem like much of an Atomic Age to me, at least at the time!

My grandfather lived through other remarkable developments like the airplane. The Wright brothers made their first flight on December 17, 1903, at Kitty Hawk, North Carolina. The flight covered 40 meters (120 feet) and lasted 12 seconds.

The first flight of the Wright brothers at Kitty Hawk, North Carolina.

Image Credit: Photographer John T. Daniels, Library of Congress
(https://www.loc.gov/item/00652085)

Initially there were just barnstormers—crazy people flying around showing off this new miracle called the airplane. They would land in open fields or were at the county fairs. You could even take a ride, if you were brave enough or crazy enough, and… had the money to pay the pilot. My grandmother did take a ride in the late twenties or early thirties. I do not know what happened, but she never flew again after that for the rest of her life. It must have been one hell of a ride, but that was the 20th century…one hell of a ride. Planes were everywhere for me as far back as I can remember.

It wasn't long before there was commercial aviation that linked every part of America and the world and soon… jets. I remember the time before jets were common. It wasn't long before you could send a letter, a package or even a person almost anywhere in no time at all. By the close of the 20th century, there were internationally over 1.4 billion airline passengers per year. The 20th century was one hell of a century. The 20th century was the age of aviation!

But then there was Sputnik (I was in high school), and then men were going into space. It was the Space Age! A man set foot on the Moon in 1969. That was a total fantasy when I was young. You heard about the possibility of these kinds of things, but they were, at the time, science fiction or fantasies. No one really believed that men would go into space and return in the 1950s when I was young. That type of idea was purely science fiction.

Did I mention computers or the internet? This was a revolution that few really appreciated either the magnitude or the incredible impact they would have on our life and work at the time.

GCOS: "What age was the 20th century…really?"

I do not know what future historians will ultimately call the 20th century. It was truly a magnificent time to be alive. My grandfather and I lived through all those changes of the late 19th and 20th centuries. The technological changes were like miracles. Man tamed the environment like no time in the past—ever. We began to change and control everything around us. Information about almost everything had expanded, and even now there is no sign that the Information Age won't continue to expand indefinitely. Perhaps we should name the 20th century after the major changes that occurred. We will be lumpers and lump all these incredible accomplishments together. They really consist of astonishing technological developments and incredible increases in information…in every field. We might call the 20th century, the century of **Information and Technology**, the IT Century. It all happened in the IT Century, the 20th century.

What will happen in the 21st century or the new millennium? We have already changed everything around us. What is left to change? What else could we possibly change?

There is one thing that we have changed very little…ourselves. We are the same old us. Yes, we are living longer, but we are the same old us. The new century, the 21st century and the new millennium may be the age when man remakes himself. Currently, we are our biggest problem. We are the world's biggest problem. We Humans are the most in need of change. Yes, we are clearly the only really "advanced intelligence" on the Earth today, but in the possible world of future advanced intelligence, current Humans will be considered…well…primitive.

Where are We (Humans) Going (as a Species) in the Future?

When we look back into the past—the journey of man revealed a fascinating reality that far exceeded any mythology ever dreamed or proposed. Reality has literally surpassed mythology.

Reality has exceeded fiction and even myth. Truth is even more powerful than myths. From many perspectives, the answers to the question of "where are we going?" are heavily dependent on the past. We live continuously in the present—which is only too soon to be the past. However, the past always leaves its mark on the future. As we have seen, the past is here today as manifested from fossilized dinosaur bones to the genes and metabolism that man shares with other animals and even ancient bacteria. The past literally has become a part of the future. However, the past itself—good, bad or indifferent is gone—never to return. The past is…history. We live continuously in the present, but we are a product of the past and move relentlessly into the future. The future is where we are going, **but what is that future for mankind?**

Probably the most important prediction for the future is that although there will assuredly be plenty of new gadgets, in reality, the most important thing that will change in the future will be mankind. During all the history of man many things have changed, but mankind has changed very little, and it has often taken tens of thousands of years for significant changes to occur in mankind! Evolution works slowly—very slowly.

Predicting the future is a very difficult task, particularly when it involves new scientific discoveries and advancements or basic evolutionary changes. However, these are exactly the future areas that we will be discussing. The joke is that you predict the future and God (the all-powerful and all-knowing God who knows the future) laughs. The "joke" reflects considerable insight into the difficulty of accurately predicting the future.

Our approach to the future of mankind will be to extend some of the current science (the usual approach). This is a common and relatively safe approach to predicting the future, but we will also liberally draw on the "perceived possible" using one of man's most important tools for significant change—science. Equally important, we will be ready to examine at least some aspects of man's social future. Mankind has advanced significantly individually, but we have also created a new entity—society, an organized, self-aware mass existence that is a function of the interaction and "intelligence" of a large group of individuals. We will be projecting what will be, as well as what should be, but perhaps also what might conceivably be possible in the future—at least in principle. I therefore offer the following chapters as a combination of what may be man's future, as well as what should or could be man's future…our **Manifest Destiny** projections are a combination of logical extensions of current technology mixed with some (hopefully) insightful intuition and undoubtedly at least a small dose of wishful thinking. Under the best of circumstances these prophecies are a combination of projected and perceived desirable changes in the basic constitution of Homo sapiens, and a blueprint for future man or whatever (whomever) is going to come after man.

On the other hand, perhaps these predictions will be only a few misguided hunches or perhaps poor science fiction. The future will decide when it fades into the reality of the past. **One of the important axioms underlying these prophecies of the future is that man will be an important player in deciding what the future will hold.** The future is no longer simply what will happen, it is what intelligent (and unfortunately sometimes not so intelligent) Humans will make happen. Humans are now in charge of our own future—or at least we will significantly shape and mold it. Whatever the mind of man can possibly conceive—if possible, in principle—is potentially achievable in our future. "Possible in principle" can roughly be translated as possible within the laws of physics.

Interestingly, even the most educated "futurists" seem to predict or assume that mankind will continue to evolve as mankind in the future. The fact is that almost no species on Earth has stopped continuing to evolve (unless they became extinct). For example, mankind is thought to have evolved from Homo erectus, which had a species existence of about 1.7 million years (I am not going to worry about or argue about possible intermediate species between Humans and Homo erectus, if any—that is for the specialists, and early relationships appears to be frequently changing). The average "lifespan" of mammalian species is about 2 million years. However, Neanderthal, who is thought to be closely related to mankind lasted only about 300,000 years (Europeans and Asians have about 4% of our genes inherited from Neanderthal, while Africans show no genetic relationship to Neanderthal). Did mankind have anything to do with the extinction of Neanderthal? Again, that is a subject (argument) for specialists, which we will not get involved with. The important concept we are looking at is that all living creatures have evolved, and most of the older species have become extinct. Mankind is probably not going to be any different. It is just a matter of time. However, we are also going to apply an increasingly recognized principal: that mankind is becoming intimately involved with our own evolution and other creatures as well, and in fact, significant manipulation and evolution of our own species is not only possible, but also highly probable. Humans will be involved with the evolution of many plants and other animals. All indications are that this process of mankind being intimately involved with evolution is not only going to continue indefinitely, but the process will probably increase dramatically in the future. In other words, **we are in the process of taking over our own evolution**; we have already manipulated and changed genes within the Human genome. Do not be concerned; we cannot possibly do any poorer job than the process of random mutations with selection of survival of the fittest has done in the past. Many people seem to legitimately worried that mankind will make mistakes that could lead to Human diseases or even death. Yes, this has already occurred, but such problems pale in comparison to the disasters of evolution in the past. This will be discussed much more below, however there are also roadblocks ahead.

To illustrate the difficulty and the success and failure of predicting the future, and why such predictions succeed or fail, we will briefly look at two previous predictions from the past. Both predictions are important in themselves, however both predictions are relevant to our current future. They both also illustrate an important principle in the business of prognostication. We will introduce both processes to gain some insight into why what appears to be a very reasonable prediction succeeds or fails.

Thomas Malthus (1766-1834) was born at Dorking, just south of London, England, and he was educated privately at home. However, his education was later continued and formalized at Cambridge, and he eventually became a curate of the Church of England. Ultimately, he became a professor of political

economics at Haileybury College—a position that he held for most of his life. His appearance and conduct have been described as that of a perfect English gentleman. We are interested in Professor Malthus because of an *Essay on the Principle of Population* that he published in 1798 in which he noted that the "population increases in a geometric ratio, while the means of subsistence increases in an arithmetic ratio."
(www.blupete.com/Literature/Biographies/Philosophy/Malthus.htm)

I will let Professor Malthus himself explain this principle in more detail.

"The number of mouths to be fed will have no limit; but the food that is to supply them cannot keep pace with the demand for it; we must come to a stop somewhere, even though each square yard, by extreme improvements in cultivation, could maintain its man. In this state of things there will be no remedy; the wholesome checks of vice and misery (which have hitherto kept this principle within bounds) will have been done away; the voice of reason will be unheard; the passions only will bear sway; famine, distress, havoc and dismay will spread around; hatred, violence, war and bloodshed will be the infallible consequence; and from the pinnacle of happiness, peace, refinement and social advantage we shall be hurled once more into a more profound abyss of misery, want, and barbarism that ever by the sole operation of the principle of population!"
(www.blupete.com/Literature/Biographies/Philosophy/Malthus.htm)

The "vice and misery" which Malthus references as the check on population expansion is war, famine, and disease. This concept regarding the consequences of population expansion has become known as the Malthusian Doctrine. It was quite controversial when it was first proposed, and it has remained contentious for the past 200-plus years. However, the prediction appears to be based on very reasonable projections of rather obvious facts at the time…we can only feed a limited number of people with the resources that are available. However, the rather dire prophecy of what would appear to be an obvious projection of the consequences of continued population growth has never occurred. Yet, the population has expanded (even exploded) considerably since his prediction in 1800. In 1800, the world population was about one billion. The world population expanded to 6.1 billion by 2000, 7 billion by 2011 (estimated by the United Nations Population Fund, or March 2012 by the U.S. Census Bureau), and it is projected to reach 10 billion people by 2050. Where is the "famine, distress, havoc and dismay…" and the "hatred, violence, war and bloodshed…" resulting from overpopulation? Make no mistake, there has been plenty of hatred, violence, war and bloodshed since Malthus' prediction, but there is only limited evidence that overpopulation per se is the major cause of such problems rather that the inability of the world's population to spread it resources more evenly around the world. These scourges (malnutrition and poverty) of mankind have been with us throughout history independent of the size of the population. There is also still starvation in the world today, but it is no worse and possibly even much less than in 1800. Malthus' prediction of doom never occurred. Simply put, Malthus was wrong.

Why was Malthus wrong? He clearly saw and defined the problem, but he did not see the incredible advances in science, technology, agriculture, and other areas (for example refrigeration, storage, and transportation (shipping) of food), use of fertilizers, etc. that prevented the ultimate disasters that he had predicted from occurring.

However, it has often been stated that the Malthusian Doctrine is still with us even today. Disaster awaits us…sometime in the near future. Was Malthus correct all along—his predictions of calamity only delayed by a few hundred years? This is of course still possible. However, there have been other advancements such as contraception, changes in attitude toward reproduction, and the simple recognition of the importance of the population problem that has resulted in reduced population expansion in most of the industrialized countries of the world. In addition, the world's geometric population growth has been predicted to level off by 2050. However, development and feeding of "third world populations" has still not become the major goal of most economically advanced nations.

The current population growth rates in North America, Europe, Japan, Australia, and New Zealand are less than 1%. The growth rates in some countries are already negative. Russia has a growth rate of -0.6% and Hungary -0.5%. Why have the less developed nations not follow the same pattern of decreased growth rates currently seen in the more developed countries. The answers are multifactorial, but to some extent they follow Malthus' principle that poverty begets poverty. Poverty, lack of education, social attitudes regarding sex and children, and particularly the inability to separate sex from procreation—the lack of understanding and application of contraception—will result in the addition of another 4 billion hungry mouths to feed by the end of the first half of the 21st century. Can the Earth's environment sustain an additional 4 billion people in the next 40 years? We shall see.

Malthus was wrong…but just barely. The factors that saved man and his world were not known at the time of Malthus. The future included such things as improvement in agriculture, the invention of birth control methods, improved education and economic development, refrigeration and transportation, fertilizers, etc. However, as we look at these factors there is but one conclusion. Mankind saved ourselves. Fortunately, these things were each done not to prove Malthus wrong, but to solve some problem or because of increased scientific knowledge. We were lucky! There was no well-considered master plan (at the time of Malthus) to save the world from overpopulation, and it did not just happen. The important point is that it was made to happen. **A plan, made by man, and implemented by mankind.** That is the key. Perhaps, at least to some extent, the intense debate that followed Malthus' prediction may have even helped prevent the catastrophes at that time and might prevent such a catastrophe in the future.

Malthus' prediction is an example of an almost obvious prediction, based on good evidence extended into the future that did not happen. However, the reason that it did not happen is that mankind literally made it not happen.

Let us consider another futuristic prediction—this one correct...in fact, even more than correct.

Gordon Moore in 1965 in an article in *Electronics* (*Electronics*, vol. 38, no. 8, April 19, 1965, fttp://download.intel.com/research/silicon/moorespaper.pdf) noted an exponential (to the base 2 (Ln_2)) growth in the number of transistors per integrated circuit and predicted that the trend would continue. This was just 4 years after the first planar integrated circuit had been produced. Moore noted; "The future of integrated electronics is the future of electronics itself. The advantages of integration will bring about a proliferation of electronics, pushing this science into many new areas. Integrated circuits will lead to such wonders as home computers..." Moore's original graph and prediction was published with the article. It showed a graph of an increase of transistors per integrated circuit from 1959 to 1965 and predicted a continuing increase to 1973. This increase and many others related to computing power has become known as Moore's Law. Moore's prediction of the impact on electronics and home computers was truly insightful. Moore had noted that "Certainly over the short term this rate can be expected to continue...if not to increase.

Over the long-term, the rate of increase is a bit more uncertain, although there is no reason to believe it will not remain nearly constant for at least 10 years." That was 1965, and the increasing density of components in integrated circuits has remained impressive for 44 years (not the 10 years Moore originally considered), and it shows no sign of abating. In fact, Moore's law has become generalized and taken on a "life of its own." It now often refers to the geometric increases in computational speed and memory along with decreased computational cost. Will it ever end? There have been repeated warnings that Moore's law cannot continue indefinitely. That certainly sounds logical. Physicists now predict that by 2030 Moore's law will run into a solid wall—the "quantum limit". This is an absolute limit set by the laws of physics—God's limit of sorts. Moore's law cannot exceed the "quantum limit" even in principle. Moore's law must fail…soon.

But not so fast—there is another concept just over the horizon called quantum computing. Many concepts in quantum mechanics are so incredible that they far exceed anything in science fiction. Perhaps quantum computing will continue to extend Moore's law. This concept is being actively pursued. Quantum computing does not try to exceed or bypass the laws of quantum physics. Instead, it uses quantum physics as a means of computation. Instead of storing a bit, a 0 or a 1 as is currently done, a quantum computer would store a superposition of both bit values—a qubit, and a function would be computed on both possibilities at the same time. It has been estimated that "… a few hundred qubits would allow one to compute a function on more possibilities than there are atoms in the visible Universe." (Landahl, A.J. *Science,* vol. 300, no. 1509, June 6, 2003.) The generalized Moore's law has taken on the character of Mark Twain (Samuel Clemens). Mark Twain is one of America's greatest literary figures. After reading his own obituary on June 2, 1897, he noted "The report of my death has been greatly exaggerated". There have been repeated predictions of the demise of Moore's law, but to paraphrase Twain they "have been greatly exaggerated," and Moore's law lives on, at least at present.

Why was Moore's prediction so accurate—even profound. Perhaps it was a clever hunch, but there is another more basic reason. Moore, who at the time of his prediction was at Fairchild Semiconductor, a division of Fairchild Camera and Instrument Corporation, went on to head the Intel Corporation, which has been a leading player in making Moore's prediction come true. In fact, many people have actively pursued the goal which Moor had predicted. Moore's law did not just "happen." It was made to happen. Moore's law is both true and profound because it became a self-fulfilling prophecy!

We should learn the basic lesson from this experience. To make accurate predictions about the future, we must utilize appropriate data that can be extrapolated, note trends, be clever, be lucky, but most importantly, to accurately predict the future, the predictions must also be something that man needs or wants, and is willing to work to make the prediction a reality (or in the case of Malthus, not a reality). The best prediction is a prediction that grows into a self-fulfilling prophecy. People must actively pursue the prediction and make it happen—not to fulfill the prediction, but because the prediction is both an attainable and desirable goal in itself. Likewise, the prediction may fail if many try to actively prevent the prediction from happening. It is almost inevitable that man will take control of the future—in particular, the future evolution of mankind! Hopefully this will be for the betterment of man and the other creatures of Earth and the Earth itself. **Our prophecies, to have a high probability of becoming a reality, must become self-fulfilling prophecies. Prophecies will not be self-fulfilling unless they are both desirable, attainable, and "Made by Man."**

Therefore, the projections regarding the future evolution of Homo sapiens presented here will presuppose man's involvement in the outcome. In fact, we will raise that concept to a basic premise that man will be intimately involved in creating the future—for good or for bad. Even the future evolution of Homo sapiens will be directed by man rather than by natural selection. Darwinian evolution is dead, and creationism never occurred. Future evolution will be based on the principle of **Intelligent Directed Evolution**. The intelligence will be mankind's. The primary tool will be science and knowledge in the broader scope, but molecular and reproductive biology, genetic engineering, genome editing, and computational science and many new fields that we have never really heard of before will also help us along the path. We are noting a new paradigm which is already well underway. **Intelligent Directed Evolution** $_{\text{Human}}$ (**IDE**$_H$) is not something that will need to be invented in the future. It is here now. The concept of man controlling our own evolution is simple noting the obvious. Natural selection, genetic mutation, and perhaps some of the other possible mechanisms of evolution that we have previously discussed have been the workhorses that have driven evolution during the past 3.8 billion years, but we stand on the brink of a revolution in evolution. With the incredible increase in genetic knowledge and reproductive biology specifically and biology, chemistry and physics with all their subspecialties in general, we are truly at the brink of a transformation which will change man not over millions or even hundreds of thousands of years as in the past, but almost overnight. We will discuss this again in much more detail in the chapter on **IDE**$_{H}$.

The Technical Aspects of Genetic Engineering (Genome Editing) of Gametes and/or Early Embryos

As noted above, we are at a critical point in biological evolution. For the past 3.7 billion years, from the beginning of life on Earth and through to the present day, evolution has been a random process with absolutely no intelligence involved. Survival has been the only "selective criteria." However, we are now at a stage of understanding the Human DNA "blueprint," and the development of technical scientific advances for "editing the Human genome" which will allow intelligent evolution of the Human species, by the Human species. These scientific advancements have opened the potential of actually controlling evolution in Humans and even in other animals and plants, including all other living organisms from single-cell organisms, bacteria, and viruses to Human. In other words, we are about to change how evolution occurs. Potentially, evolution will evolve into a rational and intelligent process rather than being a blind random mutational process.

However, there has been considerable criticism of any attempt to modify the genetics of food plants to prevent or reduce their susceptibility to insects, microbes, or environmental conditions. The result has been that farmers must continue to use toxic chemical sprays and/or suffer significant decreased crop yields due to such factors. Although at least some of the criticisms may be legitimate, and clearly there should be careful and thorough evaluation before any such genetically modified crops are ultimately released onto the market, we should not allow fear to replace science, reason, and intelligence. We must also appreciate the realities; if genetic modification cannot be done, then farmers are going to continue to use toxic chemical sprays, and we are going to continue to eat them despite the fact that we know such chemicals are toxic. In addition, there is a continuing increasing number of mouths to be fed in the world with the prospect of massive starvation always right around the corner—remember Malthus.

Unfortunately, hysteria, rather than reason and scientific evaluation, has been repeatedly applied to any genetic modification of the Human genome of any sorts. In other words, genetic engineering itself rather than any particular genetic modification has become the enemy. The genetic modification of Human embryos has repeatedly been referred to by even well-known scientists in the field of genetic engineering as a "line that researchers should never cross." In other words, Humanity should accept the current load of genetic diseases, cancer, aging, and the limited Human lifespan (and health span), and the limited intelligence given to us by a random mutational process as the ultimate goal of evolution. The acceptance of this concept of evolution is further enhanced since many people still believe that evolution does not even exist, and that what we currently have in terms of health, lifespan, intelligence, and existence was given to us by an almighty, all-knowing, and all-powerful God (the God of Abraham) and should not be altered. We have previously addressed these issues that have no basis in fact. This is a "fairy tale" that should not be treated as facts. Genetic engineering in the future will be the "savior" of Humanity using intelligent modification of the Human genome to produce **"Intelligent Directed Evolution"** in the future with increased intelligence, decreased diseases, and increased life expectancy. If you do not think that this is absolutely necessary, I would suggest that you ask any dinosaur.

However, when we ask, "where are we going?" The real answer may not be to our liking. In chapter 3-2 *Armageddon…*, the answer to where we are going is…**nowhere**. Extinction is a very real possibility. Remember the dinosaurs; we are not immune to the same fate by any means or measure. In fact, more than 99+% of all creatures that have ever lived on Earth are now extinct. (www.gi.alaska.edu/ScienceForum/ASF7/733.html) The trilobites were a very successful species that survived for 300 million years before becoming extinct. Man has been around for a mere 200,000 years. Will Homo sapiens ultimately join the trilobites? The ultimate answer is probably…yes! But don't pack your bags just yet, there may be glorious things ahead before mankind departs.

What is the most important characteristic trait that makes man at least somewhat different from the estimated possible 3 to 111-plus million other species on Earth? It is not our upright gait, or our dexterity with our hands that really makes Humans special; it is our higher intelligence that sets us apart from all other animals and has given us this first glimpse of what is now and what might be in the future. In fact, it has taken a considerable amount of this elusive substance called intelligence parceled out to a lot of different individuals that has allowed us to begin to understand this journey from the Big Bang to Homo sapiens sapiens.

It is intelligence that has even allowed us to ask the questions: "where did we come from, why are we here and where are we going?" But if higher intelligence is the key ingredient that really defines mankind, and we are a **work in progress,** then the most important evolutionary advancement in the future will probably be some significant progression or change in this quality called intelligence. If we consider the evolutionary changes from Homo habilis, the tool maker with a cranial capacity of 590 to 690 cm^3, to Homo erectus with a cranial capacity of 800 to 1,250 cm^3 to Homo sapiens sapiens with a cranial capacity of 1,000 to 2,000 cm^3, the most important or key change has been in the Human brain and in intelligence (http://encarta.msn.com). Developing an extra set of arms by a mutation in the Hox genes so that we evolve toward the octopus could produce a new Human species, but it would not add much of a selective advantage. It also does not easily fulfill our criteria noted above of a goal men would work toward achieving. Even more important, it would not fit into the new "mechanism" of evolution by—**Intelligent Directed Evolution** $_{Human}$, rather than natural selection. The selective advantage that has been clearly driving the Hominid genus for the past several million years is increase in brain mass and function—increased intelligence. In fact, it is really intelligence that has been evolving in many animals since the Cambrian explosion. Most of the biochemical evolution occurred in the primordial ooze with single-cell microscopic creatures. Evolution since that time has only applied and refined those chemical principles. Structural principles, such as body plans (phyla), and basic structures, such as sense organs like the eye and extremities for locomotion, evolved during the Cambrian period about 520 million years ago. However, what was really the basis for the Cambrian explosion? A sense organ such as the eye is really an extension of the nervous system, and limbs for locomotion are useless or perhaps even detrimental if not controlled by a nervous system. Skeletal muscles that lose their neuronal connections will rapidly undergo atrophy. Without their connection to the nervous system, muscles simply shrivel up and die. Without a connection to the nervous system, muscles are useless. We will be discussing intelligence in detail in the chapter 3-5, The *Wise Monkey*.

With the incredible recent advancements in biological science and electronics, and especially advances in genetics, genetic engineering and computer science, it is becoming increasingly clear that evolution is moving to a new paradigm from random mutations and natural selection to **Intelligent Directed Evolution** $_{Human}$ (**IDE**$_H$). I will argue that the pace of evolution will quicken and become similar to Moore's law in electronics, which we discussed above. Remember that the generalized Moore's law predicted a doubling in computer function (roughly translated as memory capacity, speed, and processing capability) every eighteen months. In fact, Human intelligence has the potential of even more remarkable results than computer development. The reason is that we already have some of that incredibly mysterious thing called intelligence. It can simply be used to increase the intelligence Humans already have.

Tally-ho, Here We Go!

We will be looking at five possible and realistic goals for the very near future, which for our purposes here will be defined as 200 years, and one for the relatively far future, which I will consider to be a thousand years—the next millennium, 3000 B.C.E. These will almost be self-fulfilling prophecies by their very nature; however, they will have a dramatic impact on future man:

1. We will look at the process of change...**Intelligent Directed Evolution** $_{Human}$ **(IDE**$_H$**)**. We will cure disease, and we will re-engineer mankind. There will be a complete uncoupling of procreation not just from sex, but from our current concepts of parenthood. Designer babies (which is an area that opponents continually complain about), the new us through genetic engineering and reproductive biology...the Humans that we always wanted to become, will become a reality. There is nothing wrong with leveling the playing field and making all of us Bo Derek 10s.
2. Probably equally as importantly, using genetic engineering, we will extend the lifespan and health span of mankind dramatically. Aging will become a relic of the past. There is nothing, in principle, which should prevent mankind from living a thousand years and perhaps indefinitely. However, despite the fact that the foundation of aging research is being laid now, it is highly unlikely that you will wake up in the near future to find a pill on the market that will give you eternal life. This process is going to be the usual hard-fought battle for every inch of "turf" and year of life wrestled from Mother Nature and the "Great Reaper."
3. **Intelligent Directed Evolution** $_{Human}$, through genetic engineering (with all of its associated tools, including newer techniques of genome editing (to be discussed in detail later), the use of stem cells, and even the use of methods and techniques not yet discovered) will bring about incredible changes. But what will we change? After all: we (Humans) were made in the "image of God." However, the most important thing that we will change (evolve) is

our intelligence, which I will use here in the broadest sense to include social intelligence as well as the more classic characterization of sentient or abstract intelligence. There will be incredible changes in intelligence, and alteration in our reliance on emotions. In other words, a kinder, gentler Human that can live at peace with himself and with the other creatures of Earth, and more completely understand the Universe in which we live. Is that concept of the future nothing but a dream? I really do not think so, because it is something that we can and will be almost guaranteed to create. We desperately must learn how to manipulate and control our surroundings for the betterment of mankind…this would include mankind's stewardship of our environment and the Earth in general. These changes must be viewed as mandatory and urgent (not elective or convenient) if we and the Earth are to survive.

4. Additional changes in our intelligence by bio-interfacing, which will allow us to use the best of biological thinking along with the advantages in computer science, is almost guaranteed to occur (we are at the present time using bio-interfacing when we interact with a computer with a keyboard or mouse, but this process will continue to increase in efficiency and intimacy dramatically).

5. The possibility of increasing life expectancy dramatically is also certain.

6. Last, we will look at Homo sapiens when we achieve these goals. Man in the very near future—perhaps by the year 2400 B.C.E., should be evolved enough to qualify as a new subspecies, which we will call perhaps Homo sapiens Intelligenticus. Can Homo sapiens intelligentsus fulfill the manifest destiny of man? Can Homo sapiens intelligentsus eventually evolve into the one thing that man desires most: to be the Gods we have created in mythology? We (Humans), already have one foot in the God door. We are the only species that has a complex language (in fact we have thousands of them), and understands how to use abstractions, such as mathematics or building jet planes, automobiles or tall buildings and cities, or has figured out and at least primitively understands the Universe in which we live. As

we have previously conjectured, we (Humans) are the only God that has ever existed on the planet Earth. However, there is plenty of room for improvement in the future, particularly if we take over our own evolutionary process—(IDE_H).

Chapter 2

ARMAGEDDON: THE FUTURE EXTINCTION OF HOMO SAPIENS

"Some say the world will end in fire,
Some say in ice.
From what I've tasted of desire
I hold with those that favor fire.
But if it had to perish twice,
I think I know enough of hate.
To say that for destruction ice
Is also great.
And would suffice."

Robert Frost, Fire, and Ice

"And he gathered them together into a place called in the Hebrew tongue Armageddon."
Revelation 16:16, The Bible, King James Version

"…and, lo, there was a great earthquake; and the sun became black as sackcloth of hair, and the moon became as blood…"
Revelation 6:12, The Bible, King James Version

"And the stars of heaven fell unto the Earth."
Revelation 6:13, The Bible, King James Version

From poetry to prophecy, the future of Man on Earth does not appear to be pleasant.

Extinction

It is not at all certain that Homo sapiens will be around long enough to have much of a future—at least a long-term future. Extinction is the loss, death, or permanent elimination of a group of organisms such as an entire species, genus or even family. Extinct creatures are no longer living or even in existence except for fossilized remains. Extinction of species on Earth is the rule rather than the exception. It is hard to clearly see or even imagine the incredible carnage of life in the past in the mundane existence of any given modern Human lifetime. You must dig for the bones or look hard for evidence. Like most of the stops on our journey, a real understanding of the history of the fate of life on Earth is very new. They feed. They reproduce. Life goes on…but not forever. Extinction is the fate of specific life forms—this is now clear from the fossil record, but what about life in general? Will all of life on Earth eventually become extinct? The short answer is yes…it is guaranteed, but we are getting ahead of our story.

Although the dinosaurs are the most famous group of creatures that are now extinct, they are by no means an exception. The dinosaurs were some of the largest creatures ever to walk the face of the Earth, and they "ruled" the Earth for over 150 million years. They are no more. Trilobites were much smaller, but they were also a very successful group of creatures that came out of the Cambrian period 500 million years ago, and they survived for almost 300 million years. For hundreds of millions of years, trilobites were a "dominant" life-form on Earth.

Trilobite, a very successful life form with a segmented body consisting of a head, body and tail with multiple limbs, and a calcium carbonate exoskeleton. There are over 15,000 described species. The trilobites came out of the incredible explosion in the diversity of life that occurred in the Cambrian over 520 million years ago. Trilobites lived for almost 300 million years. They were one of the earliest arthropods. The arthropods are the most species-rich animal phylum living today, which includes crustaceans such as prawns and crabs and chelicerates, including spiders, scorpions, and insects.

Image Credit: Pixabay, Public Domain
(https://pixabay.com/en/arctinurus-boltoni-trilobite-fossil-883871/)

Trilobites are no more. In fact, Stefan Bengtson from the Swedish Museum of Natural History estimates that 99.99999% of all of the creatures that ever existed on Earth are no more. They are now all extinct. Why should we expect man to be any different? Oh yes…I forgot; we are special. As the sarcastic cliché says…

"We are legends in our own mind." Perhaps when we have been around as long as the trilobites or even the dinosaurs or perhaps when we are smart enough to see not only the questions and problems, but also the correct answers and solutions to the major conundrums of life and the cosmos, we may have earned the right to our current arrogant tacit assumption that we dominate the Earth, and that our species is destined to exist forever. As we have previously noted, we do not dominate the Earth either by numbers or biomass. However, if you could weigh arrogance, mankind would undoubtedly win. The fact is that despite our slightly higher intelligence, we are currently in just as vulnerable a position for disaster and extinction as were the dinosaurs or the trilobites. Therefore, the future of man may be…**none.**

Many species have become extinct because of catastrophic events on Earth. However, even without any major natural disasters, there is still a "background" level of extinction occurring every year. It has been estimated that about 10 to 100 species generally become extinct every year. This rate includes vertebrates with which we are most familiar, but also includes insects, fungi and even bacteria. (Evolution Library, http://www.pbs.org/wgbh/evolution/library/03/2/1_032_04.html)

The background rate of extinction varies between different groups of organisms. Mammals have an average species "lifespan" from origin to extinction of about one million years. However, some species may persist for ten times that "lifespan." No species has or will last forever! Extinction is (in the long term) guaranteed. This raises the question, "what about mankind?"

In addition to the rate of continuous or background extinction, natural disasters or calamities can cause a marked increase in the rate of extinction. Since the Cambrian period 540 million years ago there have been at least five major mass extinctions and perhaps as many as 20 minor mass extinctions.

The most commonly known and last of the five mass extinctions occurred 65 million years ago and is generally called the K-T event. K-T signifies that the "event" occurred at the boundary between two geologic ages—the Cretaceous and Tertiary. The "K" in K-T comes from Kreide in German, which means chalk (which describes the chalky sediment layer from that time). The K-T event is the catastrophe that eliminated the dinosaurs, and killed off 70% of the species then living, including many marine bivalves, gastropods, and foraminifera. (Sole, R.V., Newman, M. "Extinctions and Biodiversity in the Fossil Record." *Encyclopedia of Global Environmental Change.* Chichester, 2002.) Despite the loss of the dinosaurs and 70% of all then living species, the K-T extinction was one of the smallest of the five major extinctions.

In 1980, a Nobel-prize winning physicist, Louis Alvarez, his son Walter Alvarez, a geologist, and several colleagues found high concentrations of the element iridium (almost two orders of magnitude higher than "normal") in sediments from the K-T boundary. Iridium is rare on the Earth's crust but is common in the rest of the Solar System and in comets and asteroids. They proposed that the K-T mass extinction resulted from the impact of a ten kilometer-size asteroid. Later, a crater known as the Chicxulub near the Yucatan Peninsula in Mexico was identified as the impact site. (http://en.wikipedia.org/wiki/Cretaceous-Tertiary_extinction_event)

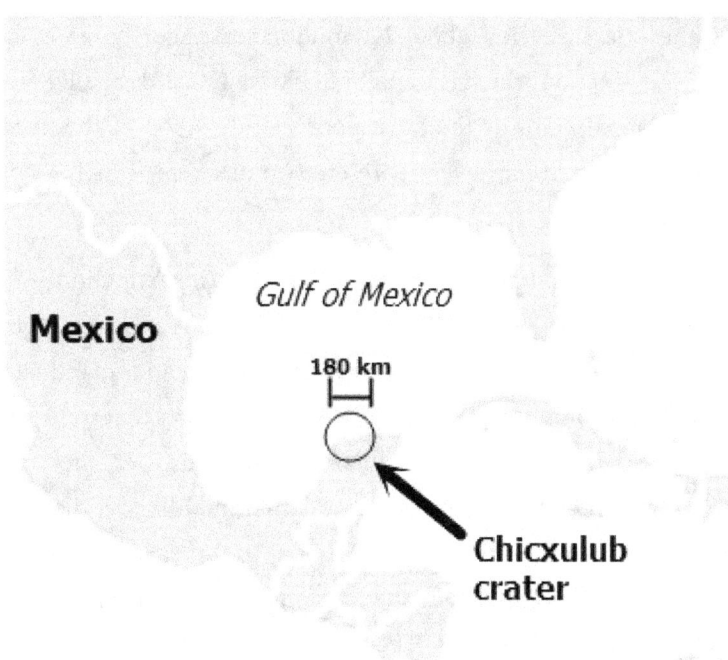

Location of the Chicxulub crater believed to be associated with the impact event 65 million years ago that killed the dinosaurs.

Image of Map Modified From: Wikimedia, Public Domain
(https://commons.wikimedia.org/ wiki/File: BlankAmericas.png)

The crater is not obvious. It is buried beneath almost 900 meters of Cenozoic sediments which have accumulated over the past 65 million years. The sediment overlies a 300-meter-thick layer of impact breccia giving unequivocal evidence of shock metamorphism like that associated with an asteroid impact. Below the breccia is melt-rock dated to 65.07 million years ago (plus or minus 0.1 million years). There is reasonable agreement amongst scientists that this is the crater resulting from an asteroid impact that was responsible for the mass extinction that killed the dinosaurs and 70% of other species on Earth at that time. (http://www.aug.org/revgeophys/claeys00/node8.html)

At the end of the Triassic period, about 210 million years ago, there was another even larger mass extinction which is thought to have killed 80% of all then-living species. This mass extinction is the most poorly understood of the five major mass extinctions. In fact, it may even represent two different "events" separated by 20 million years.

One major biological component of the end-Triassic extinction was two important groups that survived this mass extinction event and would go on to play a dominant role in life on Earth—the dinosaurs and mammals. The mammals consisted of only a group of small rat-like creatures that were probably nocturnal, lived in burrows and probably ate insects. However, they were the first true mammals. They had evolved from the mammal-like reptiles that had dominated the land for millions of years. Another group called the Sauropsids (current reptiles and birds) also flourished during the Triassic, and later in the Triassic they evolved into dinosaurs. The mammal-like reptiles were in decline even before the end-Triassic extinction event probably due to competition from the early dinosaurs. Most did not survive the end-Triassic extinction. Everything seemed set for the age of mammals (and later—Humans). However, it was not to be—at least not at that time. Early mammals were left as small furry creatures for another 150 million years until the K-T extinction when the larger and dominant dinosaurs were killed off. Dinosaurs lived up to their fierce image as aggressive competitors and dominated the Earth. Mammals were present but remained a small and insignificant group.

There are a number of theories as to the cause of the end-Triassic extinction, including the usual suspects—massive volcanism or an asteroid/comet impact. (Sole, R.V., Newman, M. "Extinctions and Biodiversity in the Fossil Record." *Encyclopedia of Global Environmental Change*. Chichester, 2002.) However, at the time, all of the continents of the Earth were massed together forming one large landmass called Pangaea.

An enormous rifting event occurred, separating the continents, and forming the Atlantic Ocean. The line of separation extended for 6,000 kilometers with an outpouring of basalt lava covering an area the size of Australia. There was also a massive release of carbon dioxide—a greenhouse gas—and sulfur dioxide (which causes acid rain).

In addition, at least two impact craters have been identified with an age corresponding to the end-Triassic extinction. One is in Western Australia and is about 75 miles (120 km) in diameter, and the other is in Quebec, Canada, and is 43 miles (70 km) in diameter. (http://palaeo.gly.bris.ac.uk/Palaeofiles/Triassic/exttheory.htm)

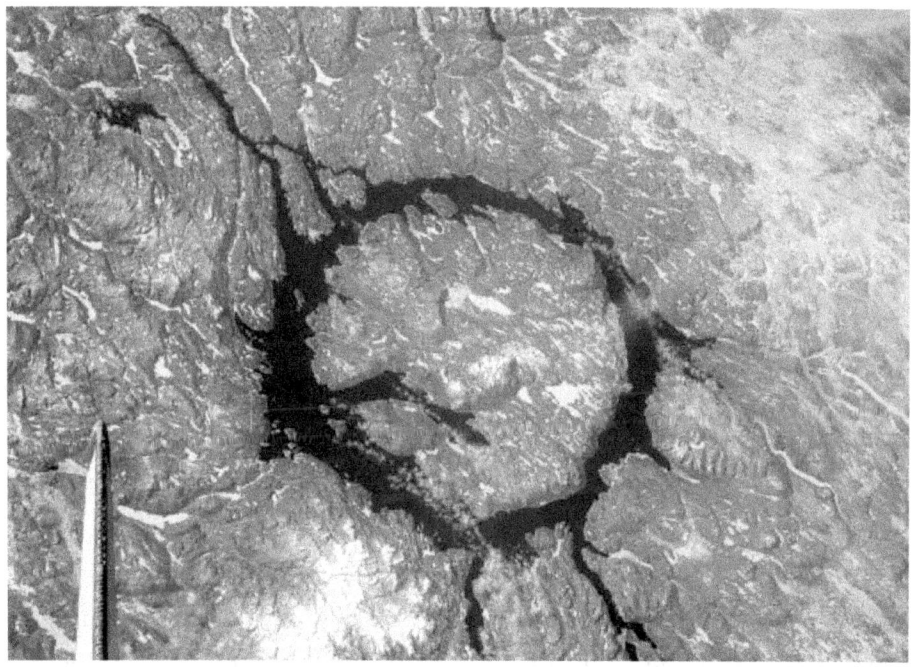

Manicouagan Impact Crater

The Manicouagan crater in Canada is 70 kilometers (43 miles) in diameter and is about 212 million years old placing it near the extinction event at the end of the Triassic. The crater has been worn down by later glaciers and erosion. The Shuttle tail can be seen on the left. The photo was taken in 1983.

Image Credit: STS-9 Crew, NASA
(https://apod.nasa.gov/
apod/ap001213.html)

Was the end-Triassic extinction due to massive volcanism and lava flows with the release of huge amounts of CO_2 gas and severe weather changes or due to one or more impact events or both? The fact is that the exact cause of the end-Triassic extinction is not clear at this time even though the body count suggests that the end of the Triassic was not a pleasant time for life on Earth.

The Permian-Triassic extinction occurred about 251 million years ago and was the most massive assault on life on Earth that has ever been recorded. About 95% of all species became extinct at that time. Marine species were particularly affected. Interestingly, the extinction event for vertebrates was much less severe than for invertebrates. There were a number of important milestones for life that occurred at the end of the Permian period. The trilobites mentioned earlier ended their long reign of existence, and the ancestors of both dinosaurs and mammals crawled out of this extinction event. During the mid-Permian period a group of mammal-like reptiles called the Therapsids had become one of the dominant land vertebrates. A primitive Therapsid called dinocephalian was up to six meters long with a meter-long head full of teeth (dinocephalian means "fearsome head"). However, this group had all but become extinct by the beginning of the late Permian period, probably due to changing climate conditions. They were succeeded by smaller Therapsids that were even more mammal-like, and possibly had fur and were perhaps warm-blooded. Some of these survived the Permian extinction, and were quite common in the early Triassic period, but were eventually replaced as the dominant land vertebrates by the dinosaurs by the end of the Triassic.

Mammals also evolved in the late Triassic period from these mammal-like reptiles (Therapsids), but as previously noted, early mammals were destined to be dominated by the dinosaurs for the next 150 million years of the Jurassic and Cretaceous period and were only minor players during that time.

(http://palaeo.gly.bris.ac.uk/Palaeofiles/Permian/vertebrates.html).

Yet, despite the incredible carnage to life ushered in by the Permian extinction, it is not at all clear what caused this epic event, and why so many different life-forms vanished from the face of the Earth...forever.

Since Alvarez's successfully proposed the asteroid theory for the K-T event, it has become popular to blame asteroids for all the mass extinctions—and with some validity. However, no crater has ever been found that correlates with the Permian extinction. In fact, it is doubtful that an impact crater will ever be found considering the significant changes in the continents and plate tectonics over the past 250 million years.

There have been reports of higher iridium content in sediments from the Permian-Triassic boundary, suggesting an extraterrestrial impact event, however the levels are at least an order of magnitude smaller than the levels found at the K-T boundary. (http://palaeo.gly.bris.ac.uk/Palaeofiles/permian/evidence.htm)

In addition to iridium, other potential impact markers have also been reported. One group led by Asish Basu, a geochemist from the University of Rochester, has reported small meteoritic fragments and microscopic bits of almost pure iron, which are thought to have condensed from the vapor of an impact cloud, in end-Permian sedimentary rock from Antarctica. (Basu, A., R., et. al. "Chondritic Meteorite Fragments Associated with the Permian-Triassic Boundary in Antarctica." *Science*, vol. 302, Nov. 21, 2003, p. 1388.) In addition, Luann Becker and others from the University of Washington have reported gas-trapping buckyballs called fullerenes containing isotope ratios of the noble gas helium consistent with an extraterrestrial origin. Fullerenes are rarely found naturally but have been found in extraterrestrial sources such as meteorites.

Fullerenes containing trapped noble gasses thought to be of extraterrestrial origin have also been found in meteoroids and from material from the Cretaceous-Tertiary boundary known to be associated with the impact event that killed the dinosaurs, but they have also been found in sediment layers from the Permian-Triassic period. (Becker, L., et. al. "Impact Event at the Permian-Triassic Boundary: Evidence from Extraterrestrial Noble Gasses in Fullerenes." *Science*, vol. 291, Feb. 23, 2001, p. 1530.)

However, at this time, like most of the other mass extinction events, it is not entirely clear whether it was a massive asteroid impact, several asteroid impacts, massive rifting events and volcanism with the release of greenhouse gases and severe weather changes or perhaps all of these occurring in some relatively short time frame (geologically speaking) that resulted in such a massive extinction event. However, it does seem to be clear that life on Earth came closer to being wiped out at the end of the Permian period than at any other time before or since. Yet we are not sure what was responsible for this catastrophic event. There are a number of smoking guns—the usual suspects—but no clear consensus on the cause. (http://palaeo.gly.bris.ac.uk/Palaeofiles/Permian/intro.html#cause)

There were two additional mass extinctions. One occurred at the end of the Devonian period 365 million years ago, and the other at the end of the Ordovician period about 440 to 450 million years ago. The Devonian extinction is poorly understood, and there is really no smoking gun.

In fact, there is not even a general consensus regarding the timing or duration of the Devonian mass extinction. The Devonian mass extinction did not have much effect on life on land but had a dramatic effect especially on warm water ecosystems, particularly the reefs. Although all the other mass extinctions appear to have occurred fairly rapidly, the Devonian mass extinction may have occurred over a period of 20 to 25 million years punctuated by 8 to 10 extinction events. (http://www.mdgekko.com/Devonian/opportunity/massExtinction.html) One theory called the "Devonian Plant Hypothesis" proposed by Thomas Algeo and others suggests that the expansion of terrestrial plants on land was the cause of the mass extinction. Plants expanding onto the land resulted in massive soil formation with the creation of calcium and magnesium carbonates and the removal of atmospheric CO_2 and global cooling. The carbonates eroded into the rivers and were exported into the oceans, forming extensive black shale deposits, which are commonly found from this period. The loss of greenhouse gases (CO_2) may have resulted in a period of glaciation.

The late Ordovician mass extinction was the second largest extinction event and occurred 440 to 450 million years ago. The cause of the Ordovician mass extinction appears to have been a severe although short-lived period of glaciation. The supercontinent called Gondwana which had been located at the equator drifted south, passing over the South Pole, resulting in widespread glaciation. Glacial deposits from this period have even been found in the Sahara Desert. Although the end-Cretaceous and other mass extinctions had a significant effect on the global ecosystem, the late Ordovician event had little permanent effect. This may be related to the post-extinction recovery process and interval. Biodiversity rebounded within about 5 million years, which is about 10 million years faster than noted after most mass extinctions. It has been suggested that the rapidity of the recovery and the lack of a profound effect on global ecosystems implies that biodiversity rebounded by immigration from other areas rather than rebounding by the production of new species as occurred in most other mass extinctions. (Patzkowsky, M. E., Pennsylvania State University, www.exopi.org/pdfs/Patzkowsky.pdf)

Many biologists think that there could be a sixth mass extinction…going on right now. We previously noted that the average "lifespan" for a mammalian species was about one million years. There are about 5,000 known mammalian species currently alive. The predicted current extinction rate should be about one mammalian species extinction every 200 years. However, in the past 400 years, 89 mammalian species have become extinct! This is 45 times the predicted rate. In addition, another 169 mammalian species are considered endangered. (http://www.pbs.org/wgbh/evolution/library/03/2/1_032_04.html)

This new mass extinction appears to be directly related to the activity of man. No asteroid, no massive volcanic eruptions, no catastrophic weather changes—just man. We are the natural disaster or calamity responsible for the Earth's sixth

mass extinction! Despite the good intentions of environmentalists and other "tree-huggers" (which includes myself), and even more radical environmental organizations, such as Greenpeace, the impact of man on the Earth's environment, and the continued extinction of species appears inevitable. It is estimated that 30% of the world's animals and plants could become extinct within the next 100 years.
(http://www.pbs.org/wgbh/evolution/library/03/2/1_032_04.html)

Unfortunately, environmentalists continue to attack the "obvious" problem of saving rainforests or natural resources rather that the root problem of Human overpopulation and poverty.

Asteroids and Periodic Extinction

Asteroids are rocky or metallic objects, most of which orbit the Sun in an area called the asteroid belt which lies between the orbits of Mars and Jupiter. There are about 40,000 asteroids that are one kilometer (0.5 miles) or larger in diameter in the asteroid belt. Ceres is the largest known asteroid and is 578 miles (930 kilometers) in diameter (about the size of the state of Texas) and was first discovered in 1801.
(http://www.enchantedlearning.com/subjects/astronomy/asteroid/)

The asteroid belt may have developed from material that never formed into a planet or from a planet that fragmented from an impact.

There are an estimated 1100 near-Earth objects (NEO) one kilometer or larger in diameter. Most of these NEOs are asteroids that have an orbit that periodically crosses the Earth's orbit or brings them into close approximation to the Earth—hence the name near-Earth objects.

THE FUTURE OF HUMANITY IS NOW

Asteroid Ida with its tiny "moon" (satellite), Dactyl (to the right).

Ida measures 56 x 24 x 21 kilometers (35 x 15 x 13 miles). Dactyl measures about 1.2 x 1.4 x 1.6 km (0.75 x 0.87 x 1 mile). Photos were taken by the Galileo spacecraft on August 28, 1993, when it came within 10,500 kilometers (6,500 miles) of Ida. Ida is heavily cratered from past small impacts. Even Dactyl would be large enough to create a major catastrophe for life if it collided with Earth.

Image Credit: NASA/JPL
(https://www.jpl.nasa.gov/spaceimages/details.php?id=PIA00069)

Some of these NEOs are all large enough to wipe out Humanity in a catastrophic impact very similar to the asteroid impact that wiped out the dinosaurs 65 million years ago. We live in a tough neighborhood. Impacts of asteroids one kilometer or larger are expected to hit the Earth about once every 100 million years. Recall that our species, Homo sapiens, is only about 150,000 years old. We have not yet as a species ever experienced an asteroid impact the size of the one that decimated the dinosaurs. Asteroid impacts are a common event in the Solar System and were even more common in the past. The cratered surface of the Moon is adequate testimony to the history of asteroid impacts in the past, and there were probably even more impacts on the Earth (because of our larger size). However, because of the Earth's active weather system and plate tectonics most of the evidence of the early asteroid impacts on the Earth has been lost. In fact, as we have previously discussed, the Moon itself is thought to have formed from a large impactor about the size of Mars hitting the Earth 4.5 billion years ago.

Do not think that asteroid impacts in our Solar System are a thing of the past. A recent example of an impact occurred in July 1994 when the Shoemaker-Levy comet broke into 21 pieces and plunged into Jupiter. Fragment G (one of the 21 fragments) has been estimated to be about two miles in diameter. The white cloud produced by the impact of this asteroid fragment on Jupiter was 14,000 kilometers in diameter with an impact core about 4,000 kilometers in diameter. For comparison, the Earth's diameter is 12,742 kilometers. The impact force of Shoemaker-Levy far exceeded the combined energy of all the nuclear warheads on Earth.

The Human race is just as susceptible to extinction by a meteoroid impact today as the dinosaurs were 65 million years ago. Had only fragment G of the Shoemaker-Levy comet hit the Earth rather than Jupiter, it would have been the dinosaur story all over again. However, this time Humanity and all larger land-based animals would probably have become extinct. Interestingly, it has taken the smartest creature on Earth 150,000 years to even realize the danger. Had an impact occurred during that time (or even still in the near future for that matter), the story of man would be at most a small footnote in the history of life on Earth that might appropriately read: "The short-lived, not-quite-smart-enough species."

What is clear is that we have only recently even become aware of one of the greatest dangers to our species and other larger animals on Earth. Ignorance is a fatal disease—ask any dinosaur.

In addition to the larger NEOs, there are an estimated 120,000 smaller impactors 140 meters to one kilometer in diameter in orbits that pass close to the Earth. Objects this size impact the Earth on average every 10,000 years. (*Science*, vol. 301, 19 Sept. 2003, pp. 1647.) Although such small impactors would probably not wipe out all of Humanity, they could have a devastating effect.

On an average night, over 100 million pieces of interplanetary debris enter the Earth's atmosphere, however most of these are small with a total weight of only a few tons. These small asteroids and comets vaporize in our atmosphere and can be seen as shooting stars. (Schweickart, R. L., Lu, E. W., Hut, P., Chapman, C. R. "The Asteroid Tugboat." *Scientific American*, Nov. 2003, pp. 54.)

Although smaller objects vaporize, larger objects explode.

On June 30, 1908, a small asteroid (or comet) about 60 meters (about 200 feet) in diameter exploded in the atmosphere over the Tunguska region of Siberia, devastating 1,600 to 2,000 square kilometers of forest. The blast had a force of about 10 megatons—big enough to devastate an area the size of metropolitan New York City. (Schweickart, R. L., Lu, E. W., Hut, P., Chapman, C. R. "The Asteroid Tugboat." *Scientific American*, Nov. 2003, pp. 54.)

The Tunguska Forest of Siberia following a blast from a 60-meter (200 foot) asteroid which exploded in the air.

Image Credit: Flickr, Public Domain
(https://www.flickr.com/photos/europeanspaceagency/42642031811)

The possibility of an impact of similar size occurring in this century is approximately 10%.
(https://www.nature.com/scientificamerican/journal/v289/n5/full/scientificamerican1103-54.html)

Fortunately, this impact occurred in such an isolated area that it took years before the full realization that an impact had occurred. Next time, we may not be so lucky.

In January 2000, a meteor two to three meters in size exploded over the Yukon Territory with a force equivalent to about five kilotons of TNT. Impacts of this size occur on average once a year.

Asteroids larger than 100 meters will penetrate deeper into the atmosphere or hit the ground with ten times the explosive force of the Tunguska asteroid. The possibility of such an occurrence in this century is about 2%.

In addition to asteroids, comets can also impact the Earth. Comets are small (1 to 40 kilometers in diameter) objects in a highly elliptical orbit around the Sun, which are composed of dust and ice as well as some rock. They have been described as "dirty snowballs." In contrast, asteroids are composed mainly of rock with little or no volatile (frozen) material. Comets account for only about 1% of the impact risks to the Earth.

We now recognize the danger of asteroid/comet impacts. It is not a matter of will there be a large asteroid impact; it is only a matter of when the next one will occur. Another one kilometer or larger asteroid impact on Earth is not only inevitable, but such an impact could also occur at any time. Although the **average** impact interval for asteroids larger than one kilometer with the Earth is 100 million years, and the last major impact occurred 65 million years ago, that does not mean the next impact will occur 35 million years from now. Four theoretical impacts with separation intervals of 150, 50, 65 and 135 million years have an average interval between impacts of 100 million years, however one impact occurred only 50 and another 65 million years apart.

In 1998, the Congress of the United States became aware of the potential problem, and Congress has mandated that the National Aeronautics and Space Administration (NASA) should find, map the orbit, and characterize the composition of 90% of these 1100 NEOs larger than one kilometer in diameter. Congress funded $4 million dollars a year to reach this goal, which was achieved in 2010. A second program with the goal of finding and mapping 90% of NEOs that are larger than 460 feet is now underway. (https://www.space.com/40239-near-earth-asteroid-detection-space-telescope.html) However, it will take many more years to map a significant percentage of these smaller, but still very dangerous asteroids.

In fact, all of these calculations are not just a bunch of theoretical abstractions. An asteroid on a collision course with the Earth has already been identified. The asteroid is called 1950 DA and is expected to cross the orbit of the Earth on Saturday March 16th, 2880.
(http://english.pravda.ru/science/19/94/377/10230_asteroid.html)

Asteroid 1950 DA is about 1.1 kilometer in diameter. Asteroids the size of 1950 DA have hit the Earth 600 times since the time of the dinosaurs. A direct impact is not expected to wipe out all of Humanity, however, if it impacts the Atlantic Ocean, it would be expected to generate a tsunami up to 122 meters high. The good news is with the present calculations of the orbit, there is only about a 0.3% risk of a direct collision. However, this estimate will undoubtedly be refined as we approach the year 2880. However, until the surveys are completed, we cannot be sure that we will not be impacted by an even larger asteroid even sooner.

The hope is that given an early enough warning, a method could be developed to slightly alter the course of the object and prevent collision with the Earth. However, it is not clear at this time that anything could be done to prevent a collision, if we suddenly found a large asteroid on a collision course with the Earth. In other words, we have progressed to the point "intellectually" that we understand the danger...and even the inevitability of another eventual asteroid impact with the Earth, but we are not quite to the point of being able to resolve the problem. In fact, before the Space Age, Humans would have been a sitting duck.

Thanks to the Congress of the United States, we are even beginning to track the orbits of most of the larger NEOs. However, there is no current developed plan that could avert the danger of an impact. If astronomers were to suddenly discover a large asteroid on a collision path with the Earth, it is doubtful that any realistic actions to avoid the impact could be taken. To put it bluntly—we are at the moment...still sitting ducks. This "situation" has some interesting resemblances to the "Dilemma of Man" that we previously discussed. We are now smart enough to understand the problem, but not quite to the point that we have a "solution." With the development of the space programs in America, Russia, Europe, and other countries, we are beginning to approach a possible "solution." However, realistically, at the present time, we should not delude ourselves into believing that we could avoid a large asteroid on a collision course with Earth. Again, we are the "in-between species." We understand the threat but are still helpless to change the inevitable outcome. Our intelligence has betrayed us...we are the sacrificial lambs. The Dilemma will only be resolved when we have both the intelligence to understand the problem, but also the intelligence to respond to the threat and control or favorably modify the outcome.

The dinosaurs did not have a dilemma. They were too dumb to understand the problem. We understand the problem...finally! The dinosaurs could do nothing about the asteroid impact that relinquished their lot to the fossil record. Neither could we—at least at the present time. Oh yeah...we would shoot a rocket at the asteroid. There have been suggestions that a nuclear blast could be utilized to alter the course of an asteroid. Yet, others have pointed out that this would only result in fragmenting the asteroid with multiple impacts that could be as bad or even worse than a single impact. Maybe we would be lucky. Maybe we wouldn't. The fact is that the smartest creature in the Universe (or at least on Earth) has only recently discovered the danger (as noted above, in the record time of only about 150,000 years, since the advent of Homo sapiens), and we are now just beginning to think about solutions. We will leave it to the reader to consider the implications of this fact on the current level of Human intelligence, which will be discussed in more detail in the chapter *The Wise Monkey*.

Fortunately, there has been some recent serious consideration of the problem. Probably the current best approach to diverting a large potential asteroid impact has been dubbed the Asteroid Tugboat or the B612 mission (named after the asteroid in the children's book, the Little Prince by Antoine de St. Exupery). The Tugboat concept has the advantage of control compared to the use of a nuclear explosion or simply smashing a large spacecraft into an asteroid at as high of a velocity as possible to divert its orbit. Asteroids or comets are often an agglomeration of material that is easily fragmented. Again, the example of this is the Shoemaker-Levy comet that broke into 21 pieces before it plunged into Jupiter.

B612 is a "plan" that is beginning to consider some of the technical problems in trying to deflect a 200-meter asteroid which would weigh about 10 billion kilograms. The concept would be for a rocket to rendezvous with the asteroid perhaps 10 years before the possible Earth impact, and gently push the asteroid off course—at least enough to miss the Earth. Although this might be a routine mission for Star Trek, the current reality of a rendezvous of a rocket with an asteroid, a soft landing and the ability to gently push the asteroid off course over a period of perhaps 10 years is above the limits of current technology. The idea would be to change the asteroid's velocity (speed it up or slow it down), which would alter its orbital period. For example, the Earth has an average orbital speed of 29.8 kilometers per second, and a diameter of 12,800 kilometers. Therefore, it takes the Earth 215 seconds to move half of its diameter.

An Asteroid heading for the center of the Earth would have to have its orbital period altered (slowed or increased) by more than 215 seconds to avoid an impact with the Earth. (Schweickart, R. L., Lu, E. T., Hut, P., Chapman, C.R. "The Asteroid Tugboat." *Scientific American*, Nov. 2003, p. 54.) A Tugboat rocket applying a one centimeter per second velocity change to an asteroid with an orbital period of two years would increase the asteroid's orbital period by 45 seconds and result in an orbital delay of 225 seconds over a ten-year period—enough to avoid an impact. However, the B612 mission has noted that using current standard chemical rocket engines, the amount of fuel necessary to accomplish such a mission would require dozens of heavy-lift rockets to boost the components and fuel into low Earth orbit, and the spacecraft would have to be assembled in orbit. It is unlikely that current chemical rocket engines are up to the task required. The plan is to develop ion engines that are more efficient. Ion engines use electrical or magnetic fields to accelerate ions out of the exhaust nozzle of the rocket. Unfortunately, ion engines also have much less thrust than current chemically fueled rockets. Most electric engines produce less than 0.1 newton of force. (Note: a newton is the metric unit of force. It is the force necessary to accelerate a 1-kilogram object at a rate of one meter per second per second). The B612 mission team has noted that 2.5 newtons are approximately the force required to hold up a glass of milk. Twenty-five ion engines would have to be ganged together to produce such force. Applying such a force to a 200-meter-wide asteroid would require 50 years to affect an adequate change in its orbital period to prevent impact. Currently the only technology that could steadily supply power for several years and could be applied to such a spaceship would be nuclear fission.

Despite their current limitations, the development of ion engines continues. NASA is currently developing a "safe" nuclear reactor to power ion engines. The Deep Space 1 spacecraft was launched in October 1998 on a mission to test advanced technologies including an ion propulsion system. The ion engine operated for 677 days but used only 72 kilograms (159 pounds) of xenon propellant. Deep Space 1 also utilized an autonomous navigation system capable of independent decision-making utilizing an artificial intelligence program called Remote Agent. The system functioned on its own, and made course corrections and detailed mission plans, and was capable of self-diagnosis without interference by ground control. Remote Agent guided the craft to the asteroid, Braille, in July 1999, and to a close encounter with the comet, Borrelli, on September 22, 2001. This mission demonstrated both the feasibility of ion engines and the ability to navigate distant asteroids and comets. However, despite its success, we are still a long way from having the capability of successfully diverting a potential Earth impactor. Clearly, we could not, at the present time, even handle a 200-meter-wide asteroid impactor, let alone a massive one- or two-kilometer impactor. However, Deep Space 1, and the B612 mission and the current program to identify potential impactors are some positive developments. We now recognize that we live in a tough neighborhood—life on Earth is no Sunday school picnic. Solving problems requires at least two processes: first, a clear recognition of the problem with its different aspects and facets, and then actively seeking solutions within the framework of the plausible (which generally means within the limits of physics). Humanity has now defined many problems and threats that could wipe mankind off the face of the Earth, including impactors from space. Solutions to the problems…we are working on them. Recall…we are a **work in progress**. The question is: "Will mankind evolve fast enough to replace arrogance with higher levels of intelligence that can recognize the problems and find real solutions, or will we go the way of the dinosaurs?"

Snowball Earth/Greenhouse Hell

Weather changes on Earth have been implicated in several of the mass extinction events, which we have already discussed. Although it has recently become apparent that life can survive in much greater temperature extremes than previously thought—such as in Antarctica or in boiling hot water, such as Yellowstone National Park or hot ocean vents—extreme temperatures still pose a limitation on life processes, and both freezing or a runaway greenhouse effect with searing temperature increases could have a considerable impact on future life on Earth—including Human life.

Probably the clearest mass extinction event associated with weather changes was the end-Ordovician extinction associated with massive glaciations, which we have discussed. However, freezing temperatures with extensive glaciations are not rare events on Earth. In fact, there have been four periods of major glaciations during the past billion years on Earth, including the end-Ordovician, during the Pennsylvanian and Permian between 250 and 350 million years ago, and the "whopper" glaciation period that occurred 600 to 700 million years ago. Another period of repeated glacial advancements and retreats has occurred during "modern times" (geologically speaking) during the past several million years. (http://www.museum.state.il.us/exhibits/ice_ages/when_ice_ages.html)

Deep marine core samples suggest that there may have been as many as 30 glacial-interglacial cycles during the past 2.6 million years. (http://www.geo.umass.edu/quaternary/pleistocene.html) During the height of the last ice age about 21,000 years ago glaciers over 2 kilometers thick covered much of North America and Europe, and sea levels dropped by 120 meters. Land and sea ice covered 30% of the Earth's surface. It has even been suggested that since the interglacial periods during these repeated glaciation cycles were longer than the period from the end of the last "ice age"—about 11,000 years ago, we may currently only be in another interglacial period with the possibility of a return of the glaciers sometime in the near future. Think of Chicago and New York under 2 kilometers of ice. That is cold even for Chicago. (I lived in Chicago for five years and enjoyed it immensely on both days of summer.) The fact is that there have been repeated periods of glaciations with interglacial periods in the recent geological past—so this possibility is not science fiction.

However, others have noted that we are currently pumping so much carbon dioxide into the atmosphere, especially from the burning of fossil fuels, that another period of glaciation could not occur. Ironically, most ecologists have been worried about the release of greenhouse gasses and a runaway elevation of temperature and a "Greenhouse Hell," while the reality may be that elevated levels of carbon dioxide may prevent us from plunging back into another cycle of glaciation. However, calculations based on the amount of solar radiation received in northern latitudes where ice sheets have formed in the past suggest that the amount of solar radiation should gradually increase over the next 25,000 years, and that no decline in solar radiation sufficient to trigger another ice age is expected for another 100,000 years, and these estimates do not include the effect of current increase in greenhouse gasses being pumped into the atmosphere which should further decrease the risk. (Hollan, Jan. http://amper.ped.muni.cz/gw/articles/html.format/orb_forc.html)

Either scenario, cold or hot, could be catastrophic for life in general, and either scenario would have a profound impact on man.

However, the reality is that we do not currently know enough about why the repeated cycles of glaciation have occurred in the past or if it would even be possible to prevent or modify such changes. Although current estimates do not predict another ice age in the near future (100,000-plus-years), we need to retain a healthy skepticism. The mere fact that there have been 20 to perhaps 30 ice ages during the past two million years or so should raise questions that another ice age is coming. The theory that the ice ages have suddenly "disappeared" just because mankind is now on the Earth smacks of Human self-centered arrogance similar to the idea that the Earth is at the center of the Universe, and the Sun circles the Earth. Whenever the answer to a scientific question seems to support the anthropic principle (the Universe was created for man)—We need to better understand the forces involved and look closely at the data for other possibilities.

If we were even considering trying to prevent a new ice age, it would require heating (or preventing cooling) of the entire Earth. Not an easy task, and not something that is within the realm of present-day technology. So, we are now literally at the mercy of the weather...just as we and all other life-forms have always been in the past!

The period of glaciation that occurred 600 to perhaps 700 million years ago is more controversial, but also may be the "Mother of all Glaciations." It has been termed "Snowball Earth."

In 1964, Brian Harland of Cambridge first suggested that there had been an early ice age after noting glacial deposits widely distributed over every continent. The possibility of a totally frozen Earth was further suggested by Mikhail Budko from the Leningrad Geophysical Observatory. Working with climate models and simple mathematical formulae, Budko showed that as the Earth's climate cooled, ice would form at lower and lower latitudes. Ice reflects more of the solar radiation than does water or land surfaces. Once ice formed below a critical latitude of about 30 degrees (equal to about half the Earth's surface area), a runaway freezing process would occur, eventually freezing over the entire planet in as little as a few thousand years. Even the oceans at the equator would be frozen to a depth of hundreds of meters to perhaps even several kilometers.

Initially the theory was not taken seriously because of the simple fact that once frozen, the icy surface would reflect so much of the solar radiation that the Earth would remain frozen solid…forever. There would be no way to melt the ice. Clearly that did not happen. However, in the late 1980s Joe Kirsch ink from the California Institute of Technology noted that volcanism would have continued even if the planet was covered in ice—what he called "Snowball Earth." Volcanism would provide large amounts of carbon dioxide into the atmosphere. Carbon dioxide is a greenhouse gas that would increase the absorption of solar radiation, warming the planet. It has been estimated that carbon dioxide levels about 350 times the present level would overcome the cooling effect of the ice. (Hoffman, P. F., Schrag, D. P.,
http://www-eps.harvard.edu/people/faculty/hoffman/snowball_paper.html)

Considering current rates of volcanic carbon dioxide emissions, it has been estimated that the "Snowball Earth" would have lasted for millions to tens of millions of years. However, once the carbon dioxide levels were sufficient to reverse the freezing, the extreme greenhouse atmosphere would have continued to have driven temperatures rapidly upward. Tropical sea surface temperatures may have reached 50 degrees Celsius (122° F). Life would have been subjected to a freeze-fry effect.

The effect of such a total freeze for such long periods of time and then the quick "fry" on life would have been extreme. However, since life 700 million years ago likely consisted of only bacteria, archaebacteria, eukaryotes, and simple metazoans, it is much harder to fully judge the extent of damage to life compared to most of the other mass extinctions that have been studied. However, the effect on the number of living organisms must have been incredible.

The most fundamental factor controlling the Earth's climate is the amount of solar radiation that the Earth receives, and how that radiation interacts with the surface of the Earth and the atmosphere. The Earth receives about 343 watts per square meter of radiation from the Sun. Some of this solar radiation is reflected into space by the atmosphere and the surface of the Earth. However, about 2/3 of the radiation is absorbed—the Earth is heated. The more solar radiation that is reflected into space, the less radiation that is absorbed. The measure of how much solar radiation is reflected into space is called the surface albedo. Snow has a high albedo (about 0.8), and more radiation is reflected and less adsorbed. Seawater has a low albedo (about 0.1), and more solar radiation is absorbed, heating the Earth. Land surfaces are intermediate between ice and water depending on the type and distribution of vegetation. When large areas of the Earth are covered by ice, the high albedo results in a large amount of solar radiation reaching the Earth to be reflected into space, further cooling the planet, and increasing the propensity for glaciation. This is the effect that Budko utilized to predict that when ice reaches the critical latitude of 30 degrees from the equator the process become self-sustaining to the extent that so little solar radiation is being absorbed, the entire Earth freezes and becomes "Snowball Earth," to use the descriptive term by Joe Kirsch ink.

However, there are other factors affecting the amount of solar radiation that the Earth receives besides albedo, and which have an important impact on global weather and glaciation. For example, the intensity of the Sun was approximately 6% weaker 600 to 700 million years ago, making the Earth even more susceptible to glaciation at that time.

In addition, the Earth's orbit, axis and change in position of the North Pole (precession) can have an important effect on weather and glaciation. The Earth's axis is currently pointing toward the North Star (Polaris) at an angle of 23.45°, and the Sun is approximately 91 to 94 million miles away. The tilted Earth revolves around the Sun on an elliptical (oval) orbit. However, both the tilt and the eccentricity of the orbit change with time—generally long periods of time.

Currently, the Earth's orbit is only slightly elliptical—it is almost a circle. There is presently only a 6% difference in the distance between the Earth and the Sun when the Earth is at its closest point to the Sun (called the perihelion), and when it is at its farthest distance from the Sun (called the aphelion). The perihelion currently occurs on January 3, and at that point on the planetary orbit the Earth is 91.4 million miles from the Sun. In contrast, the aphelion occurs on July 4 when the Earth is 94.5 million miles from the Sun. Eccentricity or change in the elliptical shape of the Earth's orbit around the Sun varies over a 95,000-year cycle with as much as a 30% difference between the closest and farthest point in the orbit. The amount of change in solar radiation due to eccentricity of the Earth's orbit is only about 0.2%, however, that is enough to have an impact on expanding or melting of ice sheets. (Illinois State Museum, http://www.museum.state.il.us/exhibits/ice_ages/why_glaciations1.html, and Matt Rosenberg, http://geography.about.com/library/weekly/aa121498.htm)

Contrary to popular belief, it is the tilt of the Earth's axis, and not the distance from the Sun that is responsible for the seasonal changes of summer and winter. The planetary axis remains fixed in space. This produces a change in the distribution of solar radiation impacting the Northern and Southern hemispheres during summer and winter. In June, the Northern Hemisphere receives more solar radiation, and is warmer, while the Southern Hemisphere receives less radiation, and is cooler. In December, the Southern Hemisphere receives more solar radiation. This can be more clearly appreciated by viewing the diagram and noting that the Earth's axis remains pointed in a fixed direction in space.

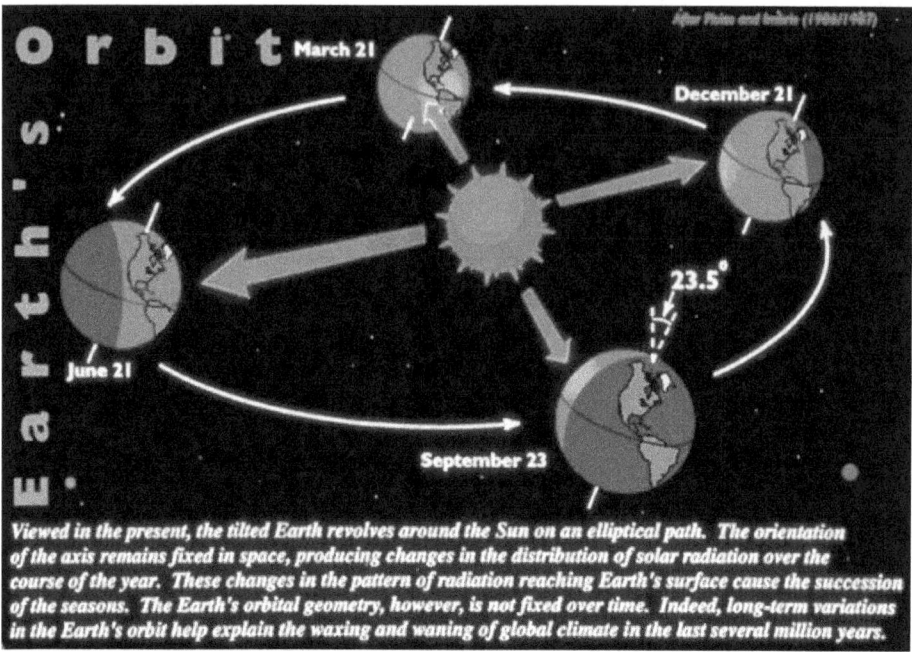

Diagram demonstrating the elliptical orbit of the Earth around the Sun.

The Earth is actually 6% closer to the Sun in winter in the northern hemisphere (at perihelion or point of the Earth orbit when we are closest to the Sun) than in June (the aphelion when the Earth is farthest from the Sun). However, the tilt of the Earth's axis, which points in the same direction in space as the Earth orbits the Sun, results in more solar radiation in June (summer) to the northern hemisphere and less radiation (winter) to the southern hemisphere. In December, the southern hemisphere receives more solar radiation (summer). However, both the eccentricity and the tilt of the Earth changes over time.

Image Credit: NOAA Paleoclimatology Program
(http://www.ngdc.noaa.gov/paleo/slides/slideset/11/index.html)

The Earth's axis also changes over a 41,000-year cycle and varies from 22.1° to 24.5° When the angle of tilt is low (22.1°), polar regions receive less solar radiation. Changes in the tilt of the Earth's axis cause marked changes in the amount of solar radiation at high latitudes. The Earth's axis also "wobbles" like a spinning top so that the North Pole describes a circle in space over thousands of years. This "wobble" or precession has a periodicity of about 22,000 years. In other words, every 22,000 years the Earth's axis completes a 360-degree circle. (Note: actually, there are two periods of 19,000 and 23,000 years with an average of about 22,000 years.) Twice a year, the Sun is positioned directly over the equator. These are called the equinoxes and occur on about March 21 and September 21. Because of the precession (wobble), both the timing of the equinoxes and the aphelion and perihelion change. Variation in the tilt of the Earth's axis and the precession of the equinoxes have a very significant effect on the amount of solar radiation (up to a 15% change) reaching high latitudes on the Earth, and greatly influencing the growth or melting of ice sheets. Currently, the Earth's axis points toward the North Star or Polaris, however, 12,000 years from now, the Earth's axis will point toward the star Vega, and the Northern Hemisphere will experience summer in December and winter in June.

The fact that the Earth's orbital dynamics had an impact on climate change and glaciation was first proposed in the mid-nineteenth century (J. A. Adhemar in 1842, and James Croll in 1864). However, initial theories and calculations did not match well with estimates of past ice ages, and the theories were discarded. However, in 1912–1914, the Serbian mathematician Milutin Milankovitch revived the orbital theory of climate change. His initial calculations and theory were largely ignored until a 1976 study of deep-sea sediment cores, which found that Milankovitch's theory and calculations correspond to climate changes and ice ages. (Hays, J. D., Imbrie, John, Shackleton, N. J. "Variations in the Earth's Orbit: Pacemaker of the Ice Ages." *Science*, vol. 194, no. 4270, Dec. 10, 1976, pp. 1121–1132.)

Interestingly, this study noted that "A model of future climate based on the observed orbital-climate relationships, but ignoring anthropogenic effects, predicts that the long-term trends over the next several thousand years is toward extensive Northern Hemisphere glaciation." It appears that changes in the Earth's orbit, axis and position of the North Pole (precession) are major factors in determining long-term climate changes, and that these changes are cyclic. Will there be another ice age…bet on it. Could mankind survive another "Snowball Earth" that lasted a million years?

The Earth's tilt is an important component impacting on the surface temperature. The Earth's axis "wobbles" like a spinning top, so that the North Pole describes a circle in space.

When the tilt is minimal, the polar regions receive less solar radiation. Currently the Earth's axis is at 23.5 degrees. When the tilt is at a minimum, polar regions receive less sunlight, when it increases, polar regions receive more sunlight.

Perhaps we are preventing another ice age with our current global warming. However, the interaction of the many different effects is complex, and it is doubtful that we are going to be able to control factors such as the Earth's tilt any time soon.

What about the possibility of a "Greenhouse Hell" on Earth? During the early period of Earth's existence from about 4.6 to 3.9 billion years ago, the Earth was probably as close to the classic concept of Hell than at any other time since, but this early period on Earth was not truly a "Greenhouse Hell". As we have previously discussed, this period of the Earth's existence is geologically called Hadean. It was so hot that the Earth's surface consisted of molten rock. However, the intense heat during this period was the result of the processes involved in the formation of the Earth per se and included the heat generated by the collision of large micro-planets and the continued intense asteroid bombardment of the Earth as well as gravitational heating and some internal radioactivity. There was probably also some contribution from high concentrations of methane, carbon dioxide and other greenhouse gases in the atmosphere at that time, but greenhouse gasses were not the major factor.

On the other hand, the high concentrations of atmospheric carbon dioxide which has been suggested to have ended the "Snowball Earth" and resulted in a very hot or "fry" period following the "big freeze" is perhaps an example of a real "Greenhouse Hell." Although there have been repeated warnings and expressions of concern from some scientists that modern industrial practices and the release of carbon dioxide from the burning of fossil fuels is pushing the Earth toward a "Greenhouse Hell," the carbon dioxide concentration during the "fry" period has been estimated to have been at least 350 times the present atmospheric concentration. (Hoffman, P. F., Schrag, D.P., http://www-eps.harvard.edu/people/faculty/hoffman/snowball_paper.html) So what is all the hullabaloo about greenhouse gases and global warming? This is a very controversial and even emotional issue for environmentalists…as perhaps it should be. It does appear clear currently that the activities of man have increased the levels of greenhouse gasses such as Co_2 in the atmosphere. In addition, there has been a warming trend of about one degree Fahrenheit since the 19th century. This warming trend appears to be associated with a decrease in snow cover in the Northern Hemisphere and floating ice in the Arctic Ocean. Globally, sea levels have risen 4 to 8 inches.

(http://yosemite.epa.gov/oar/globalwarming.nsf/content/Climate.html)

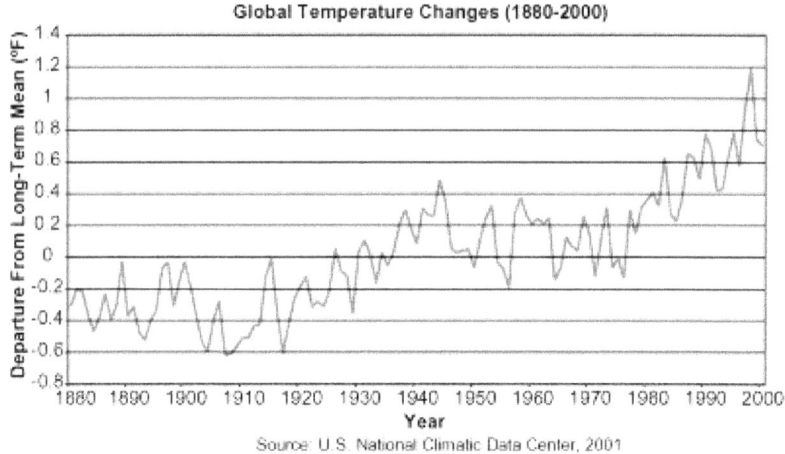

Image Credit: U. S. National Climatic Data Center
(http://yosemite.epa.gov/oar/globalwarming.nsf/content/Climate.html)

The Intergovernmental Panel on Climate Change has projected a further global warming of 2.2° to 10°F by the year 2100. The range results from the uncertainty in the levels of greenhouse gas emissions. However, even the lower end of the range would be the greatest rise in temperature seen in the past 10,000 years. The Earth's average temperature is 60°F, and a global rise of 2 to 10 degrees could have a considerable impact. Evaporation would increase, which would trigger increased rain, and intense thunderstorms would become more frequent. Sea levels could rise two feet. However, that is still a long way from a "Greenhouse Hell" scenario. On the other hand, greenhouse gases, and many other factors affecting climate change, such as Milankovitch cycles, require continued study and monitoring. Considering the past record, the Earth has been very resilient over geologic time scales in terms of rebounding from extremes of climate conditions—both freeze and fry. Conversely, it is not nearly as clear that Humans could survive a "Snowball Earth" scenario for 10 million years (or even that we could survive 10 million years in a much less harsh environment).

Sun Supernova and the End of the Solar System

Stars have a "life cycle." They are born, "live," and then die. Our Sun was born about 4.6 billion years ago when hydrogen began to fuse into helium. Our Sun will die about 5 billion years from now, when it has exhausted its fuel supply. As we have previously noted, the fuel that powers the Sun currently is the conversion of hydrogen into helium. The Sun is an average star. It is similar to 95% of the stars that make up the Milky Way Galaxy. The "death" of these types of stars is very common. In fact, the death of stars is inevitable—it is "built into the system." In the Milky Way Galaxy, a Sun-like star dies roughly every month. When a star runs out of hydrogen fuel, which has been converted into almost pure helium, the core contracts and heats up even more. The increased heat will begin to convert the helium into carbon with the release of even more heat. The amount of light emitted from the Sun will increase by a thousand-fold. The increased heat will cause the outer portions of the Sun to swell. The inflated dying star is then called a red giant. When our Sun begins to swell and become a red giant, its surface will extend past the orbit of Venus and may even engulf the Earth. If the red giant extends far enough to engulf the Earth, the Earth will be vaporized—do I need to mention that Homo sapiens and all life and everything else on Earth will also be vaporized! If the red giant stage of the Sun does not totally engulf the Earth, the Earth will simply be burnt to a cinder—do I need to… Over the following hundred thousand years, the outer parts of the Sun will be blown off into space forming what is called a planetary nebula. The term is a misnomer. Planetary nebulae were named by Sir William Herschel, an 18th century English aristocrat with a propensity for astronomy and a large telescope (for the time) in his backyard. These dying stars appeared to him as vague cloud-like objects—planetary nebulae. The name—planetary nebula—has stuck despite the fact that this process has nothing to do with planets (other than vaporizing those planets nearby). (http://www.astro.Washington.edu/balick/WFPC2/) About 40% of the Sun's mass will be blown into space. The remaining core of the Sun will then remain as a burned-out white dwarf about the size of the Earth. It will have a density of a ton per teaspoon. The 5% of the remaining stars that have masses eight times larger than the Sun will also ultimately end their lives, but as a supernova.

The ejected material from planetary nebulae consists of heavier elements "cooked" in the stars' core. The heavy elements, such as carbon, are ejected into interstellar space and may serve as raw material for future generations of stars and planets, and even future life. It is past planetary nebulae from former dying stars that we can thank for providing the carbon and other heavy elements necessary for life on Earth. We have literally been born from the death of stars that have preceded us. However, we will also inevitably pay back the piper—our turn to enrich deep space with heavy elements like carbon will come.

The beautiful but often bizarre and sometimes symmetrical shapes of many planetary nebulae that have only recently been appreciated from Hubble telescope images reveal that the process of planetary nebula formation is not a simple explosion as once thought, and the details of the processes involved are not as well understood as previously envisioned.

Planetary nebula formation by the Sun is inevitable...It is built into the physics. However, the "big event" is billions of years in the future, and there will be many other catastrophes that man will have to face before this final finale.

Man is Doomed

Considering the extinction rates of all mammals and even all known species on Earth, and particularly considering the incredible potential dangers Humanity (and all other life forms) faces, it is reasonable to conclude that man will not survive—at least in the long-term. Asteroid impact, volcanism, climate changes with another "Snowball Earth" or formation of another Venus-like Earth from greenhouse gasses or the end of the Solar System as a planetary nebula leads to the inevitable conclusion...man will not survive. It is only a matter of time—our extinction on Earth is inevitable and guaranteed. Like most of our "brothers in life" before us, we will become extinct or at least extinct on Earth!

The religious zealots carrying signs announcing the end of the world— **"The End Is Near"** ... may be right. On the other hand, "the end is near" may not occur for millions or even billions of years—so don't leave just yet, the game isn't over. To quote the great American philosopher Yogi Berra, "It ain't over 'til it's over." (http://www.yogi-berra.com/yogiisms.html) We will not survive—Armageddon is our fate, but there will still be a lot of interesting baseball before the final inning.

Chapter 3
INTELLIGENT DIRECTED EVOLUTION BY HUMANS (IDE$_H$)

Man, the Remaker of Man

Nobelist George Wald wrote, "Recombinant DNA technology (genetic engineering) faces our society with problems unprecedented not only in the history of science, but of life on Earth. It places in Human hands the capacity to redesign living organisms, the products of some three billion years of evolution." Genetic engineering or genome editing is the process of altering genes to produce a new phenotype (the observable physical characteristics of an organism). In other words, a new us! What is even more profound is that this new power will be under mankind's control, and it will no longer be a random process with significant "progress" (beneficial changes) requiring millions of years, and in addition, with the sacrifice of myriads of individuals along the way who receive a disastrous mutation that not only did not advance either the individual or the species in general, but produced a disease or impairment or shortening of the individual's lifespan. In other words, Darwinian evolution, from the point of view of individuals, and even each species has been a catastrophic disastrous process. We have in the past simply accepted evolution and a blind, unintelligent random mutational process, because there was no alternative. Similarly, we have accepted aging and death because we have had no choice. Scientific advancements in understanding the process of heredity and evolution are going to allow mankind to intelligently direct future evolution, and the potential benefits will be stunning!

With the new process of **Intelligent Directed Evolution by Humans (IDE$_H$)**, rapid progress can be made in a matter of weeks. We will discuss this process in detail below. **Implied in this paradigm, evolution will be for the first time intelligently directed.** Concern for the wellbeing of the individuals involved as well as the species will be paramount. This concern for the individual imbedded in **Intelligent Directed Evolution by Humans (IDE$_H$)** is a paradigm change with enormous ramifications. Mankind is rapidly approaching this critical point of evolution by IDE, where man will be the re-maker of mankind. Man is currently in the process of taking control of our own evolution, and the evolution of many other creatures around us…particularly domestic animals and plants. This should be recognized as a profound paradigm change in the basic evolutionary process itself.

In fact, it may surprise you, but this process (in at least a very limited fashion) has been going on for some time. Selective breeding was used in the past to produce new breeds of cattle, sheep, chickens, and most other domestic animals with desirable phenotypes (physical properties) as well as crops, such as rice, corn, wheat, etc.

Many new breeds of dogs have been produced by selective breeding. Can you believe that a chihuahua and a bichon frise are descendants of a wolf? Thoroughbreds are a horse breed which was developed in 18th century England using about 35 English mares and three Arabian stallions. They are the fastest of all horse breeds and can achieve speeds of just under 40 miles per hour for distances of a mile. Selective breeding to achieve some particular phenotype (speed in the case of thoroughbreds) is nothing more than a "low tech" form of genetic engineering. The first chapter in Darwin's *Origin of Species* is titled *Variation Under Domestication,* in which Darwin discussed the domestic breeding of new breeds by selection of specific desirable variant phenotypes. Darwin even became an avid pigeon breeder and joined two of London's pigeon clubs to learn more about the breeding of variations of pigeon phenotypes (like pigeon beaks) and the breeding of other selective phenotypes. As we have previously noted, phenotypic variation with the addition of natural selection of particular variations is really the basis of Darwin's theory of evolution. Natural selection simply means that the environment, along with competition for food, survival and mates, does the selection rather than a breeder. Natural selection is the selection of variants generally over long periods of time. On the other hand, selective breeding (with a Human selecting and propagating the desired traits) can produce phenotypic changes much faster. However, with what is now being termed genetic engineering (genome editing), dramatic changes will be done in incredibly short periods of time compared to the time scale of evolution in the past or even compared to selective breeding. In addition, it will be possible to select which specific traits (genes) are to be modified, deleted or entirely new genes inserted. With knowledge of the genome's base sequence and individual genes and their function, precise gene manipulation is now possible. In fact, there are already gene libraries with myriads of Humans, animal, plant, fungal, protozoan, and bacterial gene sequences known. We do not even know the function of many of these genes at this time.

When we looked at mankind and his kin, we noted that the genetic changes that had occurred since man and chimpanzees shared the last common ancestor was only about 0.8% (in coding regions) of the genome (DNA), but those changes had taken 6 million years to occur and has resulted in dramatic differences between the two species. It is estimated that the mutation rate of DNA is one point mutation per 2.2 billion nucleotides per generation—very slow. However, small changes can result in dramatic differences as is evident comparing chimpanzees and mankind. (http://home.att.net/~pdeitik/overview.html) Considering that the Human genome contains about three billion nucleotides, the spontaneous mutation rate would translate into only one or two new nucleotide changes per generation (in the germ line). Of course, you would also inherit some of the mutations that occurred in your mother and father and grandparents, etc. Evolution, when viewed through the actual chemical and biological processes brought about by spontaneous random mutations, is a very slow and (as we shall discuss) painful process (for the individuals involved). That is why significant "positive or advantageous" evolutionary changes often take millions of years to occur, and what results is selected positively or negatively by survival and not necessarily by their "advancement" of a species.

In contrast, with genetic engineering, significant changes could be noted almost immediately, and those changes could be widely disseminated into the Human population in as short a time as a few generations. In addition, the "changes" could be targeted to produce marked advancement in intelligence, health, lifespan, and other desirable traits (Note: I am not going to argue about desirable traits of the Human species at this time, but assume that they will be defined, and their value debated in the future). However, we will be discussing four specific areas of future genetic evolution (IDE_H), which are almost guaranteed to be in the playbook of future evolution: elimination of diseases, increasing lifespan and healthspan, increasing intelligence, and perhaps the bionification of future mankind (roughly, the interfacing of Human biology with machines and computers and/or other forms of artificial intelligence—this is already being done "loosely," when we drive a car or fly in an airplane or use a computer). This bionification will probably occur in at least two areas. One at the level of individuals, and another at the level of groups of individuals (something like the internet, but much more integrated), yielding a level of intelligence at an even higher level of existence where each individual is only a small part of the overall intelligence. In other words, evolution is not going to be limited to only individuals or a species, but to groups of individuals forming a new entity composed of individuals that will be functioning as a new super "entity."

In a comparison of the processes of classic Darwinian evolution versus genetic engineering, we might evoke a metaphor of a race between a snail and a bolt of lightning. Forget about the classic race between a tortoise and a hare. We are well beyond those classic comparisons. When genetic engineering becomes fully "perfected" (or at least advanced to the point of easily modified specific genomic changes), it will transform medicine, including the treatment of cancer, eliminate or disable endogenous viruses such as HIV and Herpes simplex, and potentially eliminate almost all diseases, and even allow the modification of biologic processes such as aging. Even the whole area of infectious diseases, which despite the advances of antibiotics could potentially be affected by introducing genetic changes to the immune system to enhance responses to certain infections or directly render viruses or bacteria incapable of infection, could markedly advance with genetic engineering. Likewise, there are a whole class of diseases called autoimmune diseases, such as lupus erythematosus and rheumatoid arthritis, where the introduction of genes to better control immune responses could be curative or at least very helpful. This could potentially even make organ transplant possible without the need for immunosuppression a reality (by modifying the organ transplanted or eliminate or modify the immune response or even growing a whole needed organ in a laboratory with self-antigens, so there would be no immune rejection).

In fact, considering that most processes in the Human body are under genetic control, the possibilities are endless. In addition, genetic engineering does not necessarily have to be restricted to Human genes. There are currently studies to understand the genetics of how salamanders regenerate a lost limb. Insertion of non-Human genes is also entirely a future possibility. Again, the potential in medicine is almost limitless. I would consider genetic engineering, in it fully developed scope, to be the most incredible advancement in medicine that will make the antibiotic era, immunization, anesthesia and modern surgery, blood transfusions, and even the basic knowledge of diseases gained in the whole 20th century look primitive by comparison. However, such medical changes will pale in comparison to the incredible future advancements in intelligence, wisdom, lifespan, and idealized future Humanity phenotype in general.

In addition, the treatment of inherited genetic diseases will come within the realm of treatment, which has currently really not been possible other than by aborting a fetus, which is clearly not an acceptable treatment from the point of view of the fetus (abortion is the only current option for the "treatment" of trisomy 21 or Down's syndrome), and is clearly not an acceptable treatment from the point of view of many religions, philosophies and people in general.

Another category of evolutionary carnage that we should mention is extinction. Extinction needs to be clearly recognized as a part of the "evolutionary process." Extinction is the part of the evolutionary process that did not make it. We have already noted that greater than 99.99% of all of the creatures that have ever lived on Earth are now extinct…they no longer exist…except as fossils. If we were considering **Intelligent Design**, then almost all of the creatures that have ever lived on the Earth were poorly designed (at least for the particular time and particular environment in which they became extinct) …they did not make it. In fact, extinction is a rather obvious refutation of **Intelligent Design**, especially by an all-knowing and powerful intelligent God as the designer. If these creatures were intelligently designed, why did they all become extinct? Had they been designed a little better…a little larger or perhaps smaller, or run faster, or were a bit stronger or had a better sense of smell or better eyesight or were able to tolerate cold better…and so on and so forth for each creature that has become extinct, they would not have become extinct. The obvious fact is that the incredible majority of all creatures that have ever lived are clearly "design failures" (at least in their environment at the time), otherwise they would not have become extinct. So much for the theory of **Intelligent Design** by some great intelligent designer! Extinction is prima facie evidence that **Intelligent Design** by an omnipotent God is nonsense.

For those that want to hold on to the creationist concept of God, then God also needs to be given his due "credit" for all of this carnage…congenital anomalies, spontaneous abortions (miscarriages occur in 10–15% of all pregnancies), pain and suffering, and the ultimate affront for each individual…death, and the ultimate insult and outrage for any species is extinctions. We should also add all the diseases and frailties that mankind (and other creatures) suffers from. Most (if not all) diseases have a genetic component. You are literally dying from "poor design." If God is responsible for **Intelligent Design** and all these other things, he obviously did a very poor job, and he (she) shows much less intelligence than what is generally attributed to him (her). The fact is, we should stop blaming God(s). In fact, this rationale is just the ultimate variant of the theodicy argument. The theodicy argument raises the question of why there are "bad things," such as babies born with terrible birth defects, if there is an all-powerful and good God. The usual retort from the usual gang is: "It is all a part of God's great design that we do not understand!" When I lived in Texas, that type of statement would be considered to have a lot of similarities to the reason that cowboys wear boots. There is a lot of shit out there.

There is no evidence of **Intelligent Design** in evolutionary processes of the past, other than the fact that if you survived and passed on your genes to your offspring, then there must be something good (intelligent) in your design. However, there is considerable potential for **Intelligent Design** in the future. In addition, we desperately need an injection of some intelligence into the evolutionary process in general, and specifically considerably more intelligence into the end-product called mankind.

In addition, we need to stop trying to take the "shortcuts" to Heaven and immortality through magic and God(s). Genetic engineers (gene modification or editing) cannot possibly do any worse with **Intelligent Design Evolution** or **Intelligent Directed Evolution** $_{Human}$ than our "friend" Mother Nature (or God for creationists) has done in the past. We are now beginning to realize that all of the diseases in the world, including growing old and dying, are the results of bad genetic design. Mankind (and all creatures that have not become extinct) are a **"work in progress"** (as we have repeatedly emphasized), and we desperately need to turn the blind, pitiless, and indifferent process of classic Darwinian evolution into an intelligently directed process.

Although there is zero evidence that some intelligence (other than the laws of physics and the "forces of nature") has directed evolution in the past, it is quite probable that intelligence will help direct the evolution of living organisms and especially man and domestic animals and some plants in the future. It appears that the concepts that the **intelligent design (ID)** people have are at least partially right. Intelligence will direct evolution…**in the future**. However, the intelligence will (at least initially) be totally Human. We will refer to this as **Intelligent Directed or Intelligent Design Evolution [Human]**, because that is what it is, but we will add a subscript to clearly recognize that the intelligence is Human—**IDE$_H$**. How do we know that Human intelligence (**IDE$_H$**) will be involved in evolution in the future? Because it has already begun, and it will be such a powerful and especially efficient tool that its use will be…inevitable. Advancements in biology and particularly in genetics and reproductive biology and some of the incredible new technology in electronics and computer science will be the fuel that drives the engine of **IDE$_H$**. **IDE$_H$** will be the mechanism by which man will remake man—it is starting to happen right now before our very eyes.

It is first important to recognize that we are in the beginning of a new evolutionary paradigm. We are currently "creating" a new subspecies of Homo sapiens sapiens, perhaps to be called in the future Homo intelligenticus—the markedly advanced intelligent monkey.

Do not fear! Yea, though we walk through the valley of the shadow of death, we will fear no evil: for we cannot possible do any worse with **IDE$_H$** than Mother Nature has done in the past using the blind, pitiless and indifferent process of evolution. Every living creature that has ever existed has died from this process. The use of intelligence in selecting our own evolutionary future should markedly accelerate and improve our evolution and reduce the suffering of mankind (and hopefully other creatures as well).

Genetic engineering or what is more recently being called gene editing will undoubtedly have an impact on the aging process, possibly by slowing aging or even reversing aging. This will almost inevitably have a dramatic impact on longevity and the maximum Human lifespan. We must recognize aging as a "disease process" rather than a "natural process." We will deal with this as a major goal of **Intelligent Directed Evolution (IDE$_H$)** in the chapter *Methuselah*. We will also discuss the potential for further intellectual development, which is the "factor" that has actually made us "Human" in the first place in the chapter *The Wise Monkey*.

Mankind will literally not know what hit us (as a species). Even more startling is that major changes to the Human genome are not millions or even thousands of years in the future; they are occurring now and will accelerate during the next several decades. The next several hundred years may yield a dramatically different Human. Genetic engineering has been repeatedly used in bacteria, fungi (yeast) and eukaryotes (creatures with a nucleus, mitochondria, and other organelles), and even in many animals (multicellular creatures) and also (more controversially) in plants (genetically modified organisms—GMO plants), and it has already been used in Humans to cure genetic diseases.

This really means that we are at a critical point in the evolution of Humans, and in fact, also at a critical point in the evolution of many other creatures on the planet Earth. Darwinian evolution will of course continue in parallel, however, because of its "snail pace," and the basic mechanism of mutation and selection by natural processes, it will hardly even be noticed, certainly within the time frame being considered here of two to three hundred years. In fact, we will probably have to periodically "back mutate" to eliminate the less than optimal "spontaneous" nucleotide mutations and alleles that may accumulate in the Human genome over time as a manifestation of "natural" evolution. That is because most "natural mutations" are deleterious, and not helpful, and we will have to "back mutate" in the future to get rid of them.

I would also like to note that this concept of genetic engineering our own evolution is not some startling new incredible inspiration that I have suddenly come up with. There are literally thousands of individuals that have "seen the light" in one form or another over the past relatively few years. The "beginning" of the paradigm shift in evolution being discussed really began with science fiction and has become more specific and more realistic with advancements in science. As the field of genetics has continued to develop, the pragmatic possibility of accomplishing such a goal is becoming…obvious. In fact, using genetic engineering to control the future evolution of mankind fulfills all of the criteria necessary for a prediction to become a reality as we have previously discussed.

Genetic Engineering (IDE_H) is badly needed, highly desirable and realistically achievable.

However, what is not obvious is what the future model of Humans will evolve into or exactly how (technically) that will occur. In addition, we must recognize that what is occurring is not just some advancement in medicine such as the introduction of antibiotics, it is a whole new paradigm shift in evolution itself. Different facets of the concept can be found in everything from popular literature such as Aldous Huxley's *Brave New World* (published in 1932), which does not paint a particularly desirable picture of the future of mankind, to Lee M. Silver, a Princeton geneticist, who published *Remaking Eden* (New York: Avon Books, 1997), and Gregory Stock's book *Redesigning HUMANS* (Boston: Houghton Mifflin, 2002) to myriads of research papers, opinions or comments in genetics, biochemistry and molecular biology that each contribute an infinitesimal to the future reality of mankind controlling our own evolution. To appreciate the magnitude of the interest in this area, and the number of concepts (both positive and negative), in addition to the amount of research that is literally pouring out of laboratories around the world, a Google search for "genetic engineering" generated 7,130,000 hits, and a search for "molecular biology" generated another 47,400,000 hits. Molecular biology is simply the "downstream" life chemistry that results from genetics. In addition, there are a myriad of other categories of life biology and even psychology and sociology that could easily fit under the umbrella that will be impacted by future genetic engineering.

The potential future of the evolution of mankind has even generated new terminologies such as transhumanism, which (by one advocate's web page) is defined as "a philosophy that Humanity can, and should, strive to higher levels, both physically, mentally and socially." (http://www.aleph.se/trans/) The pathway to fulfill those goals is the use of genetic engineering to control the future of evolution, especially of Humanity. The transhumanism movement (philosophy) was founded by F. M. Esfandiary who was a very vocal advocate for a transition of mankind into the future. He legally changed his name to FM-2030, because he believed that in the year 2030, he would celebrate his 100th birthday in a "magical time" where people would be "ageless and everyone will have an excellent chance to live forever."
(http://www.lightmillennium.org/fall/fm_interviewpart1.html)

Raymond Kurzweil, formally from MIT, who is the author of several books about computers (*Age of intelligent Machines* [1990] and *Age of Spiritual Machines* [1999]), and who is a pioneer of artificial intelligence and pattern recognition technology (read as optical character recognition) has also been a leader in projecting the future. In 1976, Kurzweil coupled aspects of pattern recognition technology to a text-to-speech synthesizer to allow reading by the blind. Interestingly, Kurzweil also picked the year 2030 as a future timeline when knowledge and technologies should be developed to allow individuals to possibly live forever. Kurzweil believes that technology will progress very rapidly in the 21st century with 20,000 years of scientific progress in the 21st century instead of only 100 years of progress as assessed by the number of years per se. In other words, from the year 2000 to 2010, there would be 20 years of technical and scientific progress rather than only 10. From 2010 to 2020 there would be 40 years of technical progress rather than 10, and from 2020 to 2030 there would be 80 years of technical or scientific progress. Therefore, Kurzweil predicts 140 years of technical and scientific progress from 2,000 to the year 2030. This implies that there will be as much scientific and technical progress from the year 2,000 to 2,030 than occurred from 1860 to the year 2000. If Kurzweil is right, and we consider the incredible changes in technology and scientific understanding that occurred during the 100 years that constitutes the 20th century, we should anticipate incredible advancements by the year 2030. Looking at GenBank, which is a repository for DNA sequencing run by the National Institute of Health (NIH), which began publishing their data in 1982 to the present, the number of bases in the repository has doubled approximately every 18 months. As of 02/15/2014 there were more than 157 trillion bases in DNA from living organisms known, and over 169 million sequences. In other words, the blueprint for how to build thousands of different living creatures are now in a computer and readily available (and as we will review below, the techniques for using this information are also being developed in parallel).

These figures show the incredible rate of increase in only five years, suggesting that Kurzweil's prediction of rapid technical progress (at least in the area of genetics) seems to be accurate. Technical progress in genetics is moving very very rapidly. Genetic scientists are in the process of knowing the sequence of almost every living organism on Earth (I would probably exclude from that, the genetic basis of every bacterium and archaeon (simply because finding every bacteria or archaea on Earth will present as least as great a challenge as sequencing them).

However, it is genetic engineering that will be powering most of the advances in future medicine, biology, and the future evolution of mankind, and probably also be the major contributing factor to Kurzweil's prediction of living long enough to live forever (which we will discuss further in the chapter *Methuselah*).

Although I love both FM-2030 and Kurzweil's enthusiasm and many of the concepts that they advocated, their timetable of a "magical time" (FM-2030) and living forever (if you make it to the year 2030 (both FM-2030 and Kurzweil)) is probably wishful thinking. I would feel that we would be progressing at a very rapid rate, if there were a few thousand Human babies that had been genetically modified (mostly to cure genetic diseases) by 2030, and we had some programs in place that were seriously attempting to identify and synthesize what I call the **"Optimal Human Allelic Set"** of Human genes by 2030 (discussed below). However, by 2030, hopefully, many if not most of the basic techniques of genetic engineering should have been worked out (such as how to effectively get new genes into the genome, or "bad genes" out, how to control gene expression, and the identification and modification of the many less-than-optimal alleles (gene variants) often associated with disease (also see below)). By 2030, genetic engineering should, at least, have begun the process of seriously evolving Humankind. Hopefully, by 2030, the genes of at least one mammal (probably the mouse) will have been synthesized "de novo" using the known gene code of the mouse, which has already been sequenced. When the techniques are developed to allow these totally synthesized genes (in the form of chromosomes) to be placed in an egg whose nucleus has been removed with the development of a live animal, and it is also possible to remove or modify any gene or place a new gene at a precise location in the genome without significant off-target mutations, then genetic engineering will have arrived and the progress will be rapid and the results undoubtedly stunning. Literally with the techniques of genetic engineering currently being developed, and the knowledge of the function of most genetic pathways, it will be possible to remake man or any species by design, not by random mutations as currently practiced by Mother Nature (or God, if you prefer).

We have previously noted that the poliovirus has been synthesized from the chemical bases using the known base sequence of polio with infectious polio being produced when this sequence was placed into a living cell. In other words, a virus has been created from chemicals using the known gene sequence of polio, which is infectious, and functions in every way (behaves) chemically like "natural" polio (not that anyone particularly wants natural poliovirus around). In fact, there is a current attempt to eliminate polio from the Earth, similar to the way smallpox has already been eliminated. We have gotten close to eliminating polio, but so far, no cigar.

More recently the complete chemical synthesis, assembly and cloning of a Mycoplasma mycoides genome has been accomplished by Daniel G. Gibson and a large group of associates at the J. Craig Venter Institute. M. mycoides has a relatively small genome (that is also why polio was used, since it is a very small virus), however, that does not lessen the accomplishment, which is proof of principle that one day the Human genome can (and will) also be completely genetically engineered. M. mycoides' genome consists of 1,007,947 bases, whose sequence was previously determined. They deleted 15 genes. In addition, they added a "watermark" that consisted of the names of the researchers involved, a URL for the project (this is the only living organism that has an "address" to its own web page in its DNA), a few quotes from the literature (using a DNA "language" that they invented), and an email address.

The genome was divided into 1,100 "cassettes," each about 1080 base-pairs long with "sticky" sequences at either end to aid in assembly.

The known Mycoplasma mycoides genome consisting of 1,007,947 bases was "cut" into 1,100 pieces, with each piece about 1080 base-pairs long, was created by Gibson et al. at the J. Craig Venter Institute, using a computer program with the pieces later chemically synthesized. Assembly of a whole bacterial genome producing the first live synthetic organism was a considerable accomplishment.

The complete synthetic genome was assembled by recombination cloning in the yeast Saccharomyces cerevisiae.

The Human genome is about 3 billion bases, which is 3,000 times the size of M. mycoides. Therefore, the synthesis of the Human genome is going to be a much bigger project and may not be accomplished any time soon.

However, bacterial genes are all in one circular chromosome while, Human genes are distributed on 46 chromosomes, so the assembly of any one chromosome would "only" involve about 65 million bases (on average) per chromosome, which would be roughly 65 times the job required to synthesize M. mycoides. In addition, there is no reason to replicate the Human genome just for a proof of principle. That has already been done with M. Mycoides.

In fact, the synthesis and assembly reported by Gibson and colleagues (including J. Craig Venter of Human genome sequencing fame) despite being almost a factory assembly line process, is still a long way from synthesizing a complete Human genome. Synthesis of oligonucleotides of the size in the "cassettes" has become so commonplace and routine that synthesis of long sequences was outsourced in M. mycoides to private companies (this was also done with polio). However, the problem (addressed below) in genetically engineering the Human genome is probably not going to be just the technical problem of synthesis or assembly (which will be considerable, because of the sizes involved), but also knowing what genes and allelic variants and SNP variants to utilize. There is also going to be considerable resistance to any synthesis (and rightly so) of a Human until these techniques are well perfected, and it can be demonstrated that they are not only safe but provide considerable advantages for the baby involved. In other words, it is unlikely any Human is going to be "synthesized" (built from scratch) just to prove that it can be done any time in the near future. Instead, the synthesis and assembly and genetic changes at the level of the germline (gametes) or embryo will probably be done one gene at a time or at most a few genes at a time to eliminate diseases. We will discuss some more of the techniques used in the synthesis of M. mycoides below.

How can the Human genome be realistically edited. When I started this book, realistic intelligent directed gene editing was not possible. However, very recently techniques have been developed to allow precise editing of genomes. In 2007, a surprise defense mechanism that bacteria use to fight off viruses was discovered. This turned out to be an adaptive immune system in bacteria. Initially, no one appreciated the incredible potential of this system, particularly in editing eukaryotic DNA even in Humans. However, by 2011 and 2012 it became increasingly clear that this system called CRISPR-Cas9 was faster and easier to utilize than any gene editing system then in existence, and the whole field has literally exploded. The CRISPR-Cas9 system allows binding to specific sequence of DNA. Scientists have now developed CRISPR-Cas9 into multiple systems. One system cuts double-stranded DNA, and another binds to a particular DNA sequence inactivating it but does not cut the sequence. One of the initial concerns were off-target binding. However, recent reports have noted techniques to markedly decrease off-target binding. Using the CRISPR-Cas9 system allows gene inactivation, gene insertion, or alteration in segments of genes. The ease of use, increasing specificity and capability of this technique with its many variations may help usher in a new era of gene editing. It appears highly likely that the CRISPR-Cas9 system or some close variant will become the principal editing system of the Human genome in the future. For those interested, I would refer you to Jennifer A. Doudna and Samuel H. Sternberg's book "A Crack In Creation" for further insight to this very powerful gene editing technique. The critically important concept is that Mankind is in the process of taking over our own evolution!

Stem cell "rejuvenation therapy" and stem cell therapy to treat disease should also be recognized as a future facet of genetic engineering, but where whole cells (generally cloned) are utilized instead of single genes. Embryonic stem cells have incredible potential in this regard. We will be discussing Dolly below, the sheep who had been cloned from a gland cell from the breast. Potentially, many different cell types could be used, including skin. The basic technique involves injecting the nucleus of the cell to be cloned into an egg (oocyte), whose nucleus has been removed. The oocyte (with the new nucleus) is stimulated to divide and is later placed into the uterus and allowed to develop. Interestingly, the egg cytoplasm appears to make the differentiated cell return to the state of a fertilized egg (undifferentiated). This is a remarkable achievement, especially since for many years making a cell become undifferentiated was considered impossible (except for cancer cells, which resulted in an entirely different problem). We will be discussing Dolly and cloning later, but the important point here is to note that it is a "technique" that is a part of genetic engineering (although a very small part). However, if the cloned developing egg is only allowed to develop to a point where there are very early cells that have not differentiated (called embryonic stem cells), and these cells can be grown in vitro, they have tremendous future potential for treating disease or even in a form of replacement therapy for damaged tissue or possibly even organs. For example, Rita Preminger from the University of Texas Southwestern developed techniques to have embryonic stem cells differentiate into muscle cells, and she used these cells to treat mice with a condition resembling muscular dystrophy. (Chu, J. *Technology Review*, Jan. 22, 2008,

http://www.technologyreview.com/printer_friendly_article.aspx?id20091)

Again, we will be discussing embryonic stem cells more later, but I stop here simply to note that these techniques are developing rapidly and will become a part of our technical ability to genetically engineer mankind in the not-too-distant future.

However, the problems (that will probably prevent their development by 2030) will not be just the genetic engineering itself or technical considerations. Many genetic engineering techniques have already been done as we have noted (with more below), and also demonstrated by transgenic mice in which a particular gene has been inserted or genes have been deleted (so called "knock-out" genes). The major problem is dissecting out which genes or which gene variants (alleles) are associated with which diseases or which genes are "optimal" for a particular function. We will also have to be aware of the gene variants called alleles. An allele is a small genetic change (base sequence change) in the DNA resulting in an amino acid sequence change in the protein) compared to the most common sequence. There are thought to be over ten million allelic variants in the Human genome. Which gene or which allele is the "best gene" to utilize? Scientists will also have to identify all of the Single Nucleotide Polymorphisms (SNP, called snips). A SNP occurs when one nucleotide in the DNA has been changed. The HapMap II project, looking at European, African, Chinese, and Japanese populations has identified 10,130,871 SNP's. (http://snp.wustl.edu/SNPseek/frequencies-by-source.html) Most of the time, changing one base in the DNA will make little difference in the ultimate function of the protein produced. However, sometime a small change can be dramatic. A good example of this is sickle cell anemia. Sickle cell anemia results when a GAG codon (in the DNA) is mutated to GTG (a SNP) in the hemoglobin beta gene, resulting in a change in the amino acid glutamic acid to valine. (http://www.ornl.gov/sci/techresources/Human_Genome/posters/chromosome/hbb.shtml) In the case of sickle cell anemia, a SNP (a change in single nucleotide) results in a very serious disease (provided it occurs in both chromosomes). All SNPs result in a new gene variant and are technically an allelic variant of the gene, however, most SNPs do not result in diseases or even in significant dysfunction of a gene. In addition, allelic variants of a gene can occur by mechanisms other than SNP, such as deletions or insertions of new DNA, etc.

We had previously talked about the vitamin C gene. Humans, chimpanzees, and great apes all require exogenous vitamin C to survive. The mouse and all other mammals can make their own vitamin C. However, the mutation causing the vitamin C gene not to function is a nine-base-pair deletion in the gene. It is the same mutation in Humans, chimpanzees, and the great apes. Since the code is read three bases at a time, the deletion results in the absence of three amino acids in the protein, but that is enough to make the vitamin C protein non-functional. However, we potentially could re-insert the nine missing bases in the vitamin C gene, and we would eliminate the disease scurvy or change the single base in sickle cell disease and eliminate that disease.

Technically, the deletion in vitamin C is an allelic variant, however it is generally referred to as a pseudogene since it is non-functioning. There are thousands of other non-functioning pseudogenes in the Human genome. In other words, there are many genes of our ancient ancestors that are still present (often mutated), but non-functional in the Human genome.

We are beginning to understand the relationship of some SNP and other allelic variants to a few diseases at this time, but at the present time, this is more like simple genetic medicine, and not something that could even vaguely be considered taking control of mankind's evolution. In other words, in order to totally re-engineer the Human genome, we will have to understand the association of most (or all) SNP and allelic variants, and their potential association with disease or less than optimal metabolism or function. It will not be adequate to simply develop techniques to affect genetic changes, in addition, we will need a clear understanding of what changes will be advantageous and what changes will lead to problems or disease, and that is what is going to take some time. This does not preclude changing or re-engineering a particular gene such as the hemoglobin beta gene for sickle cell anemia, but again, that would be more of a genetic medical advancement than the concept that we are discussing of genetically re-engineering mankind with purpose of taking control of Human evolution using the tools of genetic engineering. This broader goal will require an understanding of the pathophysiology of every disease, and how every gene and every SNP interacts with each disease, and realistically that is probably not going to happen in 10 years. However, <u>IT IS GOING TO HAPPEN, COUNT ON IT!</u>

In other words, we would have to have a detailed understanding of all of biology and medicine, and their interaction with genetics. It will eventually happen, and there will probably be considerable advancements by 2030, but it is unlikely to be recognized as a dramatic change in Human evolution by 2030. **Realistically, working out all these details may take most of the 21st century.** The problem is not just knowing **how** to change or manipulate the genetics, it also involves knowing **what** to change.

There is another factor implicit in this scenario; the changes will occur piecemeal, probably with occasional "giant leaps." In addition, the first attempts will undoubtedly be used to cure diseases (and already have been done in somatic cells, which are not passed on to future generations), rather than any real attempt to control the evolution of mankind.

However, ultimately, geneticists will come up with a set of genes that I refer to as the **"Optimal Human Allelic Set."** That will be the code for the set of all the alleles that could be used to make up the ideal Human and eliminating all the hereditary diseases and all the alleles that are associated with diseases, and all of the SNPs that produce less-than-optimal gene function. However, it should be noted that using the **Optimal Human Allelic Set** of genes to make a baby would not make him/her any different than the rest of Humanity. In fact, there will most probably not be only one **Optimal Human Allelic Set**, there will probably be hundreds or even thousands. That is because there are probably going to be many variants that can be considered ideal.

All the genes and allelic variants are still Human, however, with the **Optimal Human Allelic Set(s),** you would have the best of all of the genes in the Human gene pool in one individual. Having the **Optimal Human Allelic Set(s)** of genes would not make us more than we already are, it would simply make us all that we can possibly be genetically as a Human at this stage of evolution. Clearly, there will be multiple **Optimal Human Allelic Sets** as noted above. We would probably live longer, simply because we would be healthier and not subject to diseases, such as atherosclerosis, many cancers, etc., however, we still would not live forever, and we would probably continue to age…although perhaps slower. In addition, such changes would only affect the genetic components of disease but would not necessarily affect those environmental factors associated with disease, such as bacteria, viruses, or even our diet. When the **Optimal Human Allelic Set** (and it's probably many variants) are commonplace in Humans, we will probably be smarter, but there is no reason to think that we would be any smarter than the current smartest Human, which is an IQ of about 190–200 (using current estimates of intelligence to be discussed in much more detail later in this chapter *The Wise Monkey*). In other words, developing and spreading the **Optimal Human Allelic Set(s) of genes** through the Human population is only the first step in the future evolution of mankind. It would make us "more Human," but still **only** Human.

It is not until we learn how to manipulate the gene pool in such a way as to allow us to live for a thousand years or indefinitely and can increase our intelligence many fold can we legitimately claim that we have truly evolved mankind.

However, the **Optimal Human Allelic Set(s) of genes** would also result in "leveling the playing field" and make us all a Bo Derek "Ten" or a Michael Jordan or perhaps an Albert Einstein that can play quarterback for a professional football team in the NFL or anything else "in style" at the time of their conception. It will also eliminate much of what we now call "race." You will be able to be any race or mixture of races that your parents want or no race at all. This is one of the reasons that I have come to the conclusion that race is very unimportant, because we are literally going to genetically engineer and evolve it out of mankind's future or perhaps evolve it into the best **Optimal Human Allelic Set(s) of genes** possible. The "race" of the future will be a very smart, long lived, exceptionally healthy individual who will ultimately be given all the advancement science can develop. However, ultimately, we can advance Humankind even further by borrowing genes from other species or even totally new genes for particular purposes that can be synthesized de novo. In other words, we will not necessarily be limited to the current Human genes!

We are definitely starting on the journey of human transition, but the answers and advancements are going to come in many small pieces, possibly punctuated with a few giant leaps. In fact, it is going to be "bumpy" road at the beginning. In addition, I think that it is clearly too late for most of us alive today to be much involved in major advancements. Sorry…I would like to wave the flag of futurism, but I am restrained by my genes for skepticism and reality. In fact, the current generation of Humans will probably be one of the last sacrificial generations of Humanity. We will not make it to the "promised land" …we might call those of us alive today the "Moses Generation." As you will recall from the Bible, after leading the Jews out of Egypt and through the desert for 40 years,

Moses died before ever entering the "promised land." We can see the "promised land," but it is still down the road a bit. Our generation is still wandering in the desert, and…sorry, no promised land by 2030. The journey to the "promised land" is going to be a long hard journey with no magical shortcuts. In fact, it is exactly the continued attempts of mankind to take magical shortcuts using Gods as our solution to problems (such as disease and aging) for the past several thousand years that has resulted in mankind not having already reached some of the goals possible with genetic engineering (or at least changes that would have already brought us much closer to that time or possibility).

However, even though we will not make it to the "promised land" ourselves, we are the generations to start mankind on the road to the "promised land" …into the future. FM-2030 did not make it either. He died of pancreatic cancer in 2000, thirty years short of his goal. However, true to his philosophy, he is frozen in Arizona awaiting the future (we will discuss "freezing" as an approach to immortality in the chapter *Methuselah*, however, as a preamble, I have serious reservations regarding this approach, at least using current technology. I have frozen and successfully thawed Human cells and mouse and Human oocytes in the laboratory. It is a simple technology. However, no one has ever frozen and successfully thawed a mammal, and until I see how that is accomplished…well, I told you about that skeptic gene that I inherited. This is not to say that it cannot be done, but only to note that so far, it has **not been done**. There is nothing 'in principle' that precludes freezing and thawing a mammal, but the devil is in the details, and successfully thawing (and bringing someone back to life) will probably be highly dependent on how the freezing is done, and that is why I am skeptical of being able to "thaw" the current "crop" of people who have been frozen. Unless the "proper freezing technique" has been used, it is doubtful that these people will ever be successfully "thawed." But you never know what is coming in the future.

Ray Kurzweil's approach to getting to 2030 is vitamins, herbs, alkaline water, and other supplements…many other supplements. (VanZile, J. *Life Extension*, Sept.

2005) He has been reported to be taking 250 supplements a day. (I am not criticizing the basic principle, I take about 20 vitamins, supplements, and medications a day myself). Kurzweil is a very bright guy, and a clear leader in the concepts of prognosticating the future, and I sincerely wish him the best of luck…however, he is going to need it. (However, so will the rest of us.) He has published a book with Terry Grossman M.D. entitled *Fantastic Voyage* in which the theme is "live long enough to live forever."

They make many good recommendations in their book for a healthy lifestyle, and most of their recommendations are based on pertinent research. However, no one has ever shown that the combinations that they advocate has ever made any Human live longer (except for the obvious recommendations that if you do not smoke, you do reduce your risk of lung cancer, emphysema and atherosclerosis and will statistically have a better chance of living longer, etc.). In fact, Kurzweil, whether he realizes it or not, **is the experiment…himself.** Will it work? We will have to wait to 2030 to see if the "experiment" worked, and to see if new technologies have emerged to facilitate living forever by that time. Exactly what that future technology will consist of is not clear. Kurzweil implies that the technology will revolve around nanobots, which are (wisely) only vaguely defined in his book but appear to be a host of nanometer sized robots (what would you expect from a "computer type" guy).

On the other hand, I would envision the new technologies to be centered on genetics, molecular biology, gene editing and probably stem cells (as "replacement parts"), (what would you expect from a "biology type" guy like myself). In addition, both FM-2030 and Kurzweil's prophecies of the future of mankind centered on living forever…immortality for mankind. Although extending the lifespan of man is an important goal, I prefer to see this as just one part of the larger picture of evolving mankind using genetic engineering to create a whole new species that lives longer, eliminates, or dramatically decreases and even reverses aging, but is also results in marked increases in intelligence with eliminating most of the diseases of mankind (living healthier).

THE FUTURE OF HUMANITY IS NOW

In fact, I would be willing to bet Kurzweil a bottle of one of my Grand Cruise Classy Cabernet Sauvignons against a dollar that we will not have the answers to living even a hundred and fifty years by 2030 (current maximum lifespan is about 125 years). I have a very small collection of wines, which I have had for over 40 years. I drank one bottle when my first book came out (it was a medical textbook). I drank another bottle when my second book came out (again, a medical textbook). I plan to drink one of my remaining bottles of Chateau Lafite Rothschild when this book is published. I am saving one bottle for when I die (don't ask how I am going to drink it after I die…I am still working on that problem). However, I have one other bottle for the bet. I figure that if either of us dies, I win. If I die first, I will not have to pay up (I'm dead, and the bet is off). If Kurzweil dies first, he will not pay up either, but at least I will have the satisfaction of knowing that I was right. However, if both of us live to 2030, and the technology has developed to live forever, I lose the bet…Kurzweil wins, and I will gladly give him my bottle of Chateau Lafite Rothschild, 1973. However, I gain a lot more…immortality or at least a long lifetime…so who cares. I could be wrong **once and** live with it.

I will assume that Kurzweil sincerely believes in what he is doing (after all, he is doing it to himself). However, there is no evidence that any of what he proposes will work (other than give one a healthy lifestyle, which is admirable, and can by itself give a longer healthspan and probable lifespan). However, transhumanism and many of the current futuristic approaches smack of religion. We have already tried the faith approach to longevity (in Heaven). What this area needs is science, and not religion. Remember FM-2030; he also picked the same year 2030 as to when mankind will have the technology to live forever, and as I previously noted, he is no longer with us. Few of us will make it to 110 or beyond. FM-2030 did not make it. My bet is neither Kurzweil (born in 1948) nor any of the rest of us living today (including myself, born in 1940) will live beyond the current upper limits of Human longevity of 125! Perhaps our great-great-great-grandchildren in the 22nd century will see the "promised land" …I give that a maybe.

Unfortunately, many of these approaches like freezing and herbs and vitamins rapidly degenerate into a religion where "faith" is the only thing underlying the "logic." This does not denigrate the importance of vitamins in any way. Vitamins are a vital nutrient, and taking a daily vitamin supplement is probably a very reasonable health recommendation, but there is no evidence that vitamins and herbs are going to make anyone live much longer. (I am going to have to make a couple of possible exceptions here. There is beginning to be evidence that a compound found in the skin of red grapes (and a variety of other plants) called resveratrol may act through a gene pathway involving a gene called SIR2 and may prolong lifespan. However, this is still questionable. This has been demonstrated in yeast, the worm, C. elegans, and more recently even in a fish (a vertebrate), however, this is still quite controversial, and significant improvement in the lifespan of mice has not been demonstrated (as usual, more research is needed). There are also other possible "longevity drugs" that are beginning to be, such as Rifampin, but most of the research so far has been preliminary. In addition, Rifampin can be quite toxic, so no one (sane) is recommending it as a longevity treatment at this time. The most recent "youth drug" appears to be NAD+. However, this does not appear to be anywhere near the answer to living forever or even living to age 150. There are a lot of laboratories working on this problem though. Time will tell. Another drug that has shown some possible promise is metformin. Metformin has been suggested to extend Human lifespan by about 10 or more years; however, the studies are ongoing. I am taking metformin if that tells you anything.

A Google search for "transHuman movement" generated 921,000 hits. Apparently, the movement is flourishing, and many have seen the hope for the "light of the future." That is one of the reasons that we included the "remaking of man" among the Manifest Destiny of mankind…it is literally an obvious scenario that is almost guaranteed to happen…sometime in the future. However, when, and how we get there is clearly open to considerable differences of opinion. My approach is more along the lines of a basic principle that I learned in graduate school...

"Show me the genes," and "show me the data." Religion is best reserved for the church down the street. Faith is a wonderful thing for those that have it; it can have some definite psychological advantages...better than tranquilizers, and at times perhaps better than antidepressants. However, confusing faith with reality can also be a very dangerous miscalculation. Let's say a man has absolute faith in God. In addition, he has complete faith that God will let him fly like an Angel. If the man jumps off a tall building, he will learn the clear distinction between faith and the reality of gravity. Faith is a form of wishful thinking. Noting than mankind has evolved the highest intelligence of any species on the planet Earth; faith, life after death in a Heaven, and all the other mystical and magical "doctrines" of religion are a sad commentary on current Human intelligence.

Probably, the central theme and most important concept that is basic to transHumanism, and all the rest of what has been lumped together as "futurist philosophy" (including the last section of this book that you are reading now) is the recognition that mankind "in reality" cannot totally control our own future at the present time. However, perhaps, with further research We may not have to die in the future! We may not have to grow old! I think that in the future there is a significant possibility that we can extend mankind's life span considerably. We can control our own biology and our own evolution! We can evolve into something incredibly more "Human" and hopefully more intelligent than the current variety of Homo sapiens. However, we cannot achieve any of these goals with mega-vitamins, herbs (at least none discovered so far) or faith.

To put it in the vernacular... "We are sick and tired of all this garbage that "Mother Nature" (or God) has given us, and we are going to do something about it." There is a Hell...and we should recognize that we are living in it. We certainly do not need any philosophical discussions or "proofs" of the existence of Hell. However, we are now beginning to realize that we can potentially put out the fire. True…we have only sprayed the fires of Hell with a garden hose…so far, but the tsunami is on its way. Beating our swords into plowshares, and our spears into pruning hooks is a good idea for peace (paraphrased from Isaiah 2:3-4). However, we are not going to get to the "promised land" by building more churches or taking megavitamins. If we want a Heaven, then we are going to have to build it ourselves right here on this little rock whizzing through space…remember one of our themes…**Made by Man**. No God (except the Man-God) is going to give us anything.

Take note that the **incredible advancement in future science and evolution** is not some particular genetics discovery (even the sequencing of the Human genome…as necessary and great an accomplishment as that was), **the incredible advancement is the realization (the awakening) that we can realistically take over our own evolution and "remake mankind."** The important advancement is the recognition that there is a high probability that death, aging, and most diseases can be genetically controlled, and that **it will be possible to evolve ourselves.** Despite the fact that work in many of these areas is preliminary, we will be looking at specific genes and specific data. I wish (just like everybody else) that we could be there by 2030, but the job is too formidable to be solved in just a few years even with an increased rate of research and technical development as noted above.

Intelligent Evolution

One of the most important aspects of the new paradigm in evolution by genetic engineering or gene editing and more science development is that it will be **intelligently directed.** The intelligence will be strictly Human (at least initially). This may sound a bit like **Intelligent Design (ID)** that has been recently proposed as an alternative to the classic Darwin method of evolution involving mutations and natural selection, but the concepts are entirely different. **Intelligent Design (ID)** is not a new theory of evolution at all…or a new theory of anything for that matter. ID is simply creationism under a new name that is being used for a political and legal agenda in court cases and school boards to oppose evolution. Although the major strategy of the proponents of ID try not to mention God (so that it looks like they are not discussing religion), the implications are always that the same. **Intelligent Design** poses that certain biologic processes or structures are irreducibly complex and require an intelligent designer, and the intelligence had to come from somewhere or something, and God quickly emerges one way or another as the source of the intelligence used to design whatever. **Intelligent Design** is based on what is considered irreducible complexity of some biological systems with the assertion that a certain biological structure could never have developed without a designer to show the way.

The designer is always the God of Abraham or his son Jesus. The first problem is that even if a God was required, there is no evidence and certainly no requirement that that God must be the God of Abraham, and even less evidence that Jesus is the son of the God of Abraham, who has no corporeal body, and no genes to give to Jesus (we discussed this earlier in *Man and His God's*). Of course, the answer to this conundrum is that Jesus' immaculate conception was a miracle. The next conundrum is that if an irreducible complex system required a more complex intelligence to design it or make it evolve properly, who designed or created the God with that complex intelligence? The standard answer: God created himself! Again, another miracle. When a "theory" or explanation requires miracles, it is not an explanation, it is not rational, and must be accepted for what it really is, which is nothing but babble...words without meaning.

I have already stated several fundamental principles regarding God! First, **God did not create man; man created God! Man is the God creator.** Second, **all of the thousands of Gods that have ever been defined, created or otherwise conceived in the past, have been created by mankind!** Third, **man even created all the concepts that make up the God scenario!**

However, the **Intelligent Design** proponents may have gotten it at least partially right…for the **future** of evolution. **Intelligent Directed Evolution** or **Intelligent Design** would be a much more effective and efficient technique of evolution than the current evolutionary process of mutation and natural selection by environmental factors. In fact, even the very ardent supporters of classic Darwinian evolution do not seem to fully appreciate what an appalling disaster evolution really is on the level of the individuals. This, of course, does not make classic evolution wrong, it simply points out that evolution is not a Sunday afternoon picnic, and it is not intelligent. In fact, classic evolution is a rather slow, dumb and "painful process." Evolution's only virtue is that given enough time and enough corpses along the way, what comes out may be better (survivable under any environmental circumstances) than what went in. However, there are a lot of individual creatures along the way that paid the ultimate sacrifice for evolution by random mutation.

It is truly amazing that so many people fear genetic engineering of the Human genome and oppose any effort to control the future evolution of man. What they should fear is doing nothing, because Humanity has paid and continues to pay a very high price for **not** having taken over our own evolution in the past. The current method of Darwinian evolution (or if you prefer Mother Nature or if you think God is responsible for creation…then God) results in 10–15% of all pregnancies and developing embryos to "spontaneously" abort. In other words, they die before they are ever even born. In fact, studies looking at very early pregnancies (even before there is a missed menstrual period) place the spontaneous abortion rate even higher…perhaps as high as 25% of all conceptions (pregnancies). These babies do not develop properly or die in utero (in the uterus of the mother) and are passed (aborted spontaneously). Sometimes they have developed sufficiently to look like a baby with well-formed arms and legs and a head and face; other times when they die very early, they are passed as a jumble of cells and tissue mixed with blood and placental products and decidua (part of the maternal lining of the uterus). They have formed nothing more (literally) than a bloody mess. Sometimes they are malformed. Studies of why these babies die have noted aneuploidy (abnormal numbers of chromosomes) in almost half. In other words, there are abnormalities in numbers of chromosomes or deletions (part of a chromosome is missing) or some other chromosomal abnormalities. It should be appreciated that a single chromosome is a huge amount of genetic material (average about 250 million nucleotides in length) and no one has ever been reported to have been born and lived missing a whole autosomal chromosome (all the chromosomes except the X and Y chromosome). Humans can survive missing one X chromosome, such as in Turner syndrome, with an XO karyotype and only 45 chromosomes (rather than the normal 46 with a XX (female) or XY (male) karyotype), but that is one of the few exceptions. Of course, all males have only one X chromosome, and females do not have a Y chromosome.

Most other developing embryos die because they have unfortunately inherited a defective gene necessary for cell function or embryo development from one or both parents (even though their karyotype is normal). Sometimes the mutation is not inherited but occurs "spontaneously" during fertilization or early embryo development. Although abortions can occasionally occur from infections, endocrine problems (such as severe hypothyroidism, etc.), and a host of other problems, the great majority of abortions result from abnormal genetic factors. The reproductive process for producing Humans is incredibly inefficient; however, this is classic evolution in progress. In fact, classic evolution as manifested by the incredible number of "bad experiments" reflects the Dawkins' view of the Universe (in this case, the classic Darwinian evolutionary process) as a blind, pitiless, and indifferent process, and clearly a process that shows no regard for Human life nor any consideration for any mystical "sanctity of life." However, evolution does differ somewhat from Weinberg's concept of a "Pointless Universe," which we have discussed. The Universe may be pointless (unless, as we have discussed, a sentient creature (Humans) creates their own "point"), however evolution does have a "point." The "point" is survival. Those species that missed the "point" are no longer with us…they are extinct. However, Mother Nature (or again, God if you prefer) is totally indifferent to perhaps over 19-plus-million embryos dying every year (worldwide)…year after year after year (based on a world population of over 6 billion with a birth rate [world average] of 20–22 per thousand population, and a spontaneous abortion rate of about 15% [which may be even higher as noted above, if very sensitive techniques are used to detect all very early pregnancies that abort]). However, this is Darwinian evolution at work for you. Rarely…very rarely, some of those mutations or genetic changes may prove useful and life or mankind takes a small incremental step forward. You and I are literally what have walked out of millions of years of such carnage. However, it is clear that evolutionary and the reproductive processes "designed" by Mother Nature (or again, God if you insist) is a totally blind, pitiless, and indifferent process (as per Dawkins).

It is truly ironic that the "usual gang" declares that it is immoral to experiment on Human embryos when Mother Nature (or God) does it millions of time each year…year after year, and the overwhelming majority of the "experiments" are bad experiments and the outcome is lethal. In fact, each one of us is a small evolutionary "experiment." In addition to the millions who die before they are even born, there are more millions who have been born with birth defects. About 4% of the Human population is born with a birth defect. Most of these are minor, but about 1% represent a major problem, such as Down's syndrome (mongolism, resulting from having three of the number 21 chromosomes) or a major heart defect such as a large ventricular septal defect (a hole between the ventricles of the heart), etc. Not all congenital anomalies are the result of genetic alterations and fit neatly into a category that can be considered a part of the evolutionary process of mutations or genetic changes followed by natural selection, but most do, and they are examples of the bad or failed experiments of the evolutionary process. Evolution is certainly not "kind" to life…Human or otherwise. In fact, as we have tried to emphasize, evolution shows a total disregard for the fate of the individual of every species, or considering the percentage of all species that have gone extinct (99.9+% including early Human-like species such as Neanderthal), it is clear that evolution (even if it were controlled by a God) has no consideration for any individual or species.

GCOS: "Thanks God."
When one looks at the details of the process of evolution, it is not a pretty sight. Evolution has a dark side…a very dark side. The overwhelming majority of mutations or chromosomal changes are detrimental. Clearly, scientists cannot do much worse than what God or "Mother Nature" has done in the past. The "nose count" of the carnage along the pathway of evolution speaks for itself. The use of intelligence (even if it is only Human) rather than blind random mutations will be an incredible improvement in the evolutionary process.

However, a word of caution. **True**…classic Darwinian evolution is a disaster (from the viewpoint of those creatures that have experiences its vagaries). It would be much better and certainly more Humane, if we had a wonderful, powerful creator God that created each creature individually for their habitat and "function" in life. Not to mention the fact that mankind considers ourselves to have been created as the special "chosen species" above all other creatures that have ever lived.

GCOS: "Wow! I'll vote for that. Give me a God! Evolution is a disaster!"
Unfortunately, the world, and particularly biology, is not a democracy. Look around you. It does not take a rocket scientist to figure this one out. Literally, what you see is what you get. Life never offered us a rose garden…and we definitely did not get one.

GCOS: "Drad…no candy store and no Santa Clause either. What kind of world is this anyway?"
The answer is that we live in a "real world," where the only candy stores are the ones mankind has built for himself. When you look around, if you do not like what you see…aging, death, and disease, then you should do something about it,

and that is exactly what will be done with genetic engineering resulting in **Intelligent Directed Evolution** $_{\text{Human}}$ **(IDE$_H$)** by taking control of our own evolution. Mankind has lived through (or more correctly survived through) classic Darwinian evolution for hundreds of thousands of years (actually millions to billions of years depending on where in life you begin counting). We have tried God(s), and that has not worked. Now it is time to do it ourselves. God(s) would be better, but you cannot create God(s) by wishful thinking. Wanting (or praying) for a God to solve the problem of mankind does not create a God, it only creates a myth and a delusion. If mankind wants to solve these problems, we will need to do it ourselves. Again, remember one of our themes… "**Made By Man**"; it is our only way forward. In other words, Mother Nature takes hundreds of thousands to millions of years to make significant changes, and God is not interested in change. If mankind wants "progress" or to rapidly and significantly change ourselves, we must do it ourselves. In addition, humans are the only Gods that have ever existed on Earth or any place else as far as we now know,

In fact, the discussion has raged around creationism and classic Darwinian evolution. However, creationism is a fairy tale, and classic Darwinian evolution is a nightmare (for the individual and usually even for species). Mankind desperately needs to take over our own evolution. What we currently have is a billion-year-old disaster, and mankind's only salvation is ourselves…**Made by Man**!

Germline Versus Somatic Cell Engineering

Gene alterations may be done at the level of the sperm or egg (germ cells) or zygote (fertilized egg with the diploid number of chromosomes before cell division), which will make the genetic alterations a heritable trait. Such altered genes and phenotypes would then be transmitted to future generations just like any other genes… "naturally." In addition, genetic engineering can be done in adults to treat diseases or even potentially change some phenotypes of an adult.

Although I am sympathetic to genetic changes that may be accomplished in an adult, which is the only type of genetic engineering that would be useful for myself and my generation and for everyone alive on the Earth today, it is germline genetic engineering that will ultimately change the fate of mankind and become the new non-Darwinian method of evolution…IDE_{H}.

There will probably always be a place for adult genetic engineering, and a great deal of the current research in this area is directed toward treating diseases, and is not germline engineering, and changes will not be passed onto future generations. Therefore, non-germline genetic engineering will not affect the evolution of mankind. In addition, adult somatic genetic engineering will be a technically difficult task, and (in some cases) will require changes in multiple base pairs of the 30 to 60 trillion cells that make up the Human body. That will not be a simple task.

In contrast, once genetic changes are introduced into the germline they can be transmitted and spread in the population "naturally" just as any other gene, however, in addition, they may also be inserted into fertilized embryos that do not have the desired gene (and phenotype). That will spread those genetic changes even faster and propel the evolution of mankind at an incredible rate.

In other words, adult somatic genetic engineering is for the individual, while germline genetic engineering is for the future of the Human race. Somatic cell genetic engineering should be recognized as a medical treatment, and could become a powerful tool, however, germline genetic engineering can evolve mankind…IDE_{H}. Although as stated above, somatic cell genetic engineering will revolutionize medicine, and will be the first to occur (in fact, it is already occurring).

In this thesis, we are interested in both somatic and germline genetic engineering, however, we emphasize that only germline changes will result in hereditary changes that will be used to evolve mankind. In other words, it is germ cell genetic engineering that will eventually result in the new paradigm of significant evolution of mankind.

Germline genetic engineering has often been referred to derogatorily as producing "designer babies." The fact is that this approach has the potential to cure most or even all of the hereditary genetic diseases (and there are a lot of them), make us immune or at least very resistant to many diseases, including cancer, atherosclerosis, and Alzheimer's disease, and potentially at least some bacteria and viruses, as well as increase the intelligence of Humankind, and perhaps increase the lifespan of mankind dramatically; it is hard to understand why there are so many individuals that are bitterly opposed to any genetic engineering involving the Human germline. Most of the criticisms are based on a complete lack of understanding or appreciation of the fact that almost all the current problems of mankind from cancer to heart disease to Alzheimer's are the direct results of our current evolutionary process, which is a "blind," random mutational system, which has absolutely no intelligence behind it other than the survival of the species. In addition, if one looks carefully behind most of the arguments against any Human genome editing, you will find the argument usually legitimately involves the effectiveness of the engineering, but eventually leads to God.

However, such concerns must legitimately be addressed, and they also suggest that such research must be done cautiously with considerable discussion along the way. In other words, research must be done cautiously and carefully. This will be discussed more below!

Steps to the Future

Genetic engineering of the Human germline will probably occur more in a stepwise fashion with the next several decades involving more the development of the "tools of the trade" rather than any massive attempt to incorporate (or delete) any specific gene or genes into the Human germline.

Gregory Stock of the University of California has outlined what he considers as the probable sequence that will eventually lead to genetic engineering in Humans. (Jonietz, E. "Choosing Our Children's Genetic Future, an interview with Gregory Stock." *Technical Review*, February 2003, pp. 78–79.) The first form of "genetic engineering" will utilize preimplantation genetic diagnosis (PGD), which are techniques to diagnose genetic diseases such as cystic fibrosis and a long list of other genetic diseases in embryos before the embryo is transferred back to the mother. This is already currently being done for a limited number of genetic abnormalities with at least moderate success. The first cases of PGD were initiated in the United Kingdom in the mid-1980s. By 2001, there were about 50 centers in 20 countries offering PGD for a variety of disorders. (Ouhibi, N., et al. "Preimplantation Genetic Diagnosis." *Current Women's Health Reports*, vol. 1, 2001, pp. 138–142.)

Approximately 2,500 clinical cycles of PGD have been done worldwide, resulting in 600 pregnancies and the birth of over 500 healthy children. (http://64.233.187.104/search?q=cache:Aq0KaWYQfqYJ:www.who.int/reproductive-health/infertility/21.pdf+%22preimplantation+genetic+diagnosis%22+karyotype&hl=en)

PGD is generally done by removing a polar body or a single blastomere (one cell in the early embryo usually at the eight-cell-stage of development).

A variety of techniques have been utilized to detect aneuploidy, including fluorescent in situ hybridization (FISH), which is an old technique first introduced in 1977. The FISH technique is rapid and allows a very quick diagnosis of certain types of aneuploidies like Down's syndrome and has even been used in research for identification of the location (on a chromosome) of a specific gene. However, it is relatively limited in the amount of genetic information obtained. A variety of other different techniques have been developed to detect chromosomal abnormalities. An even older technique is karyotyping in which all the chromosomes are stained with a dye and abnormalities in chromosome number of even deletions (where a piece of a chromosome is lost) or translocations (in which a piece of one chromosome is attached to another chromosome) can be identified. Karyotyping is a very important diagnostic technique for genetic diagnosis, and about 400,000 karyotype analyses are done each year in the United States and Canada (not all of these are done on fetuses in utero). However, many of these are done in association with amniocentesis (drawing amniotic fluid from around the baby to allow genetic analysis). Amniocentesis is performed much later in a pregnancy (generally 17 to 20 weeks gestation) and is not preimplantation genetic diagnosis (PGD). In fact, the whole purpose of PGD is to diagnose the abnormality **prior** to implanting the embryo, and generally not implant those embryos that have the disease. Amniocentesis and karyotyping are also old techniques (which I have been doing as a physician for 35 years). Amniocentesis involves aspirating some amniotic fluid (with a long needle) that surrounds the baby. The amniotic fluid also contains cells from the baby which are grown in culture until there are enough cells to do the karyotype analysis, which generally takes a week or two. (http://www.biology.arizona.edu/Human_bio/activities/karyotyping/karyotyping.html) However, classic karyotyping is generally not done in PGD, because current PGD techniques are working with only one or two biopsied cells and the results are needed rapidly (before the embryo is transferred into the mother). Amniocentesis also has risks for the baby with a death rate as high as 1 in 200 resulting from infection, premature labor, bleeding, or direct injury to the fetus. More recently, there have been attempts to collect cells shed from the cervix for karyotyping with less risk to the fetus. About 7 to 10 days after conception, the developing embryo separates into an outer cell mass destined to become the placenta and an inner cell mass which will become the embryo. Some of the superficial placental cells (trophoblasts) are shed into the cervical canal and can be isolated and karyotyped. (Amiel, A., et. al. "Journal of the Society for Gynecologic Investigation," vol. 12, 2005, p. S290.)

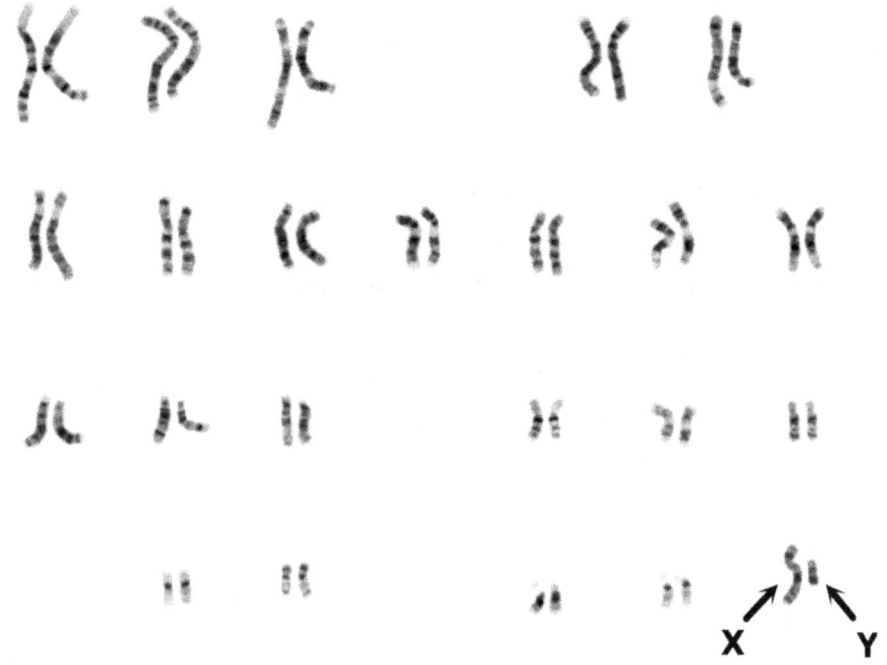

Human karyotype (male)

There are 22 pairs of autosomes plus two sex chromosomes (XX for females or XY for males). Karyotyping has been an invaluable tool for genetic analysis, particularly at the level of the whole chromosome.

Image Credit: Wikimedia Commons, Public Domain
(https://commons.wikimedia.org/wiki/File:NHGRI_Human_male_karyotype.png)

This approach would make karyotyping a much safer procedure and might make it possible to karyotype all pregnancies in the future.

In addition to karyotyping, a variety of other techniques have been used to diagnose single-gene disorders. PGD has currently been reported to have been used to diagnose 26 different single-gene disorders, and the list continues to expand.

However, techniques like **karyotyping and PGD are really rather primitive tools** and have had only a very minor (actually insignificant) impact on the Human gene pool.

Despite the current usefulness of these techniques, the ultimately diagnostic approach will be to sequence the whole fetal genome rapidly fully. However, the problem has been to obtain fetal genetic material. There was a breakthrough in 2010, when it was reported that about 12% of the maternal plasma DNA was fetal in origin, and this DNA could be sequenced. This finding, together with the marked decrease in cost, has raised the spectra that fetal whole genome sequencing may be right around the corner, which will virtually eliminate amniocentesis, karyotyping, chorionic villus sampling and all other current techniques for prenatal diagnosis and open multiple possibilities for the future of mankind.

Sequencing

The base technology that has stimulated the whole concept of genetic engineering has been DNA sequencing. Sequencing is the process of determining the order of nuclear bases that constitute the genome of any living creature. It is, in essence, the blueprint of how every organism is built, and the program that runs it.

The National Human Genome Research Institute (NHGRI) continues to coordinate efforts to dramatically reduce costs and increase the speed at which a mammalian-sized genome can be sequenced. NHGRI continues to award millions of dollars in grants each year specifically for studies to increase speed and reduce the cost of genome sequencing. Costs and time have been dramatically decreasing. The first Human genome sequencing project began in 1990 and was headed by Dr. Francis Sellers Collins using a classic sequencing technique and was sponsored by the U.S. National Institute of Health and numerous other groups around the world and involved an array of international

laboratories. As the years passed there were dramatic improvements in sequencing techniques. A separate non-government project was also instituted in 1998 by Celera Genomics, a company founded by J. Craig Venter (J. Craig Venter Institute) using a "whole genome shotgun technique." Both projects came together to independently publish preliminary draft of the Human genome in 2000, however the project was not considered finished until 2003, and by that time covered 99% of the Human genome at a cost of about 3 billion dollars. (https://en.wikipedia.org/wiki/Human_Genome_Project)
(https://www.jcvi.org/about/jventer)

In 2007, the cost to sequence the Human genome decreased to about 10 million dollars; in 2017, the cost had decreased to about 1,000 dollars. (National Human Genome Research Institute, https://www.genome.gov/27541954/dna-sequencing-costs-data/) The time needed to sequence the Human genome has also been greatly reduced. A new machine was released in 2017 by Illumina that is capable of sequencing the Human genome in one hour or less! (https://www.sandiegouniontribune.com/business/biotech/sd-me-illumina-novaseq-20170109-story.html)

It is probably very realistic to expect that by 2020, it may be possible to sequence a mammalian-sized genome in about an hour at a cost of perhaps $100.00 dollars or less. In fact, that is exactly the NHGRI's goal. If this goal is successfully met that will be one of the most incredible technical feats in the history of science. We would be firmly on Kurzweil's curve. Sequencing techniques have kept doubling their throughput every 22 months for the past 15 years. (McElheny, Victor K. http://www.the-scientist.com/2006/2/1/42/1/) One company has developed sequencing techniques that produces tens of millions of raw bases per hour on a single instrument. (http://www.454lifesciences.com/index.asp)

Genetic diagnosis will become a common medical practice, allowing "profiling" of each individual for all the diseases for which they may be particularly susceptible. This will allow the potential for intervention long before a disease ever produces symptoms, and as noted above, in the case of fetal development, whole genome analysis will allow the evaluation of which potential problem genes or alleles the fetus might inherit, with potentially allowing those problem genes to be altered, replaced, or modified. In addition, with whole gene sequencing PGD will then become commonplace; it will become the "point of contact" where Human genome engineering will occur (at least initially). Genome sequencing will pave the way for the treatment of most of the serious life-threatening genetic diseases. However, PGD will ultimately be recognized as a primitive technique that is really "too late, and too little."

That time appears to have arrived. Illumina, one of the leaders in the field of gene sequencing, announced in 2014 that the thousand-dollar Human genome has finally arrived! This is more than a six-log decrease in cost since the announcement of the completion of the Human genome sequence in 2001.

Ultimately, we will have techniques for modifying, replacing, or deleting genes almost routinely at the time of fertilization. When genetic engineering techniques are demonstrated to be safe and effective, serious but not necessarily life-threatening diseases will be treated, followed later by the selection of non-disease traits…the dreaded word again: "designer babies."

However, we need to broaden the end scenario a bit to include the selection of genes that will dramatically improve the Human genome. Although, the stepwise scenario is probably correct in at least its broad outline, the end results will not be such trivial results as "designer babies," which generally include sex selection, skin, hair and eye color, and other insignificant variants (except in composite, which we will get to momentarily).

Genetic engineering will result in the remaking of man…millions of years of evolution telescoped into a few centuries. The Human genome is packed with gene variants that function less than optimally in addition to junk DNA that does nothing (or almost nothing, but more about that shortly).

An important "first step" in genetically engineering the Human germ line has already been taken, but it was not done with any intention or consideration of genetic engineering. The technique of in vitro fertilization (IVF) was developed for the treatment of infertility. However, there can be no genetic engineering without clinical techniques to obtain oocytes (eggs), and/or zygotes (fertilized eggs) or early embryos, and that is exactly what IVF is all about. In every IVF cycle, there are (generally) multiple eggs that have been perused under a microscope by an embryologist, technician or physician, and began the development of life in an incubator in a laboratory. When the appropriate genes and the techniques for introducing those genes into the egg or zygote are developed, it will not be necessary to invent the wheel again...the clinical machinery is literally sitting there waiting for the development of genetic engineering techniques. Those techniques have already been developed in animals.

Steptoe and Edwards "conceived" the first "test tube" baby in late 1977, and Louise Brown was born on July 25, 1978. Currently, there have been over five million test tube babies born since the birth of the first IVF baby Louise Brown. Interestingly, even though some have compared the birth of the first IVF baby with a biological equivalent of the first landing of man on the Moon, the first IVF baby was not the result of any great scientific breakthrough. (Actually, the first landing of a man on the Moon is very similar in that it is hard to point to the one incredible scientific "breakthrough" that made the journey possible. Man's landing on the Moon was rather an incredible gut-wrenching inch-by-inch processing of intense engineering and most important, the realization that the trip was feasible, and even more important, the determination (and commitment of money) to accomplish that goal). IVF was similar, but the original determination was solely that of Steptoe and Edwards.

Dr. Steptoe was a gynecologist with an expertise in laparoscopy, which is a technique in common use today for visualizing and performing surgery in the female pelvis. I have been doing laparoscopies, since I was a first-year resident in 1967–68. My department had brought a professor up from Latin America (they were way ahead of us in using this technique, mainly because of less medical liability issues) who was very experienced with laparoscopy to teach interested faculty and the first-year residents the technique. Generally, a new surgical procedure would be taught to the senior residents first, but in this case, the decision was made to teach the first year's residents first, because the seniors would soon be graduating, while the first-year residents would be around long enough to learn the technique well enough to teach the other residents joining the program as the years passed. More recently general surgeons have "discovered" the technique (it only took them 25 years), and adopted laparoscopy for doing gallbladder surgery, appendectomies and even hernia repairs. Laparoscopy is now a very routine surgical technique. This is only of historic interest, because most IVF cycles are now done using ultrasound-guided needle aspiration of oocytes. However, historically laparoscopy was an integral part of the development of IVF.

Laparoscopy allowed Dr. Steptoe to aspirate oocytes (eggs) directly from the ovary. Doctor Edwards was an expert on laboratory embryo research and in vitro culture. Their "big breakthrough" in the development of IVF was Steptoe and Edwards' persistence and determination despite considerable criticism from the usual crowd for "experimenting" with Human eggs. The critics were numerous and vocal. The critics (either then or now) never seem to have appreciated the fact that without the development of the IVF technique, the millions of oocytes that have turned into living Humans through IVF during the past 36 years would have simply degenerated. Where is the Human dignity and respect for life in allowing oocytes to degenerate? Where is the "moral high ground" in **not** creating a Human life? In a real sense, had the critics been successful in stopping the "experiments," they would have, in essence, "killed" the millions of people who have been born using IVF, and all those who will be born in the future. Unfortunately, protest continues by these groups who refer to themselves as "pro-lifers," while in fact they should be called "pro-deathers." Although critics often refer to the sanctity of life as one of the reasons for their criticism of such research, the real sanctity of life was in the experiments that led to successful IVF, and not in their opposition. There is no sanctity of life to allowing an egg to degenerate in a woman who is actively seeking and wants to have a child. This is another example of paradoxical morals (morals which are touted to be good, but are in fact completely immoral), based on some misguided interpretation of what God wants. Do you hear the prophetic ventriloquists? The prophetic ventriloquists are people giving their own opinion, but using God as a dummy, while they pull the "string" to work "God's" mouth to give "God's opinion."

Steptoe and Edwards began their collaboration in 1966, and Louise Brown was born 12 years later. Edwards won the Nobel Prize in 2010. A Vatican official condemned the move as "completely out of order."

After, Steptoe and Edwards demonstrated the feasibility of IVF, the technique rapidly spread around the world. The first American baby was born in Norfolk on December 28 of 1981. IVF programs sprang up everywhere in the United States and around the world.

I participated in an early "In Vitro Fertilization Program" (IVF) as an assistant professor at Baylor College of Medicine beginning in 1981 (planning), and finally a clinical active program in 1982–3, and I later started another IVF program myself at East Tennessee State University in Johnson City Tennessee, where we had a set of twins on our third IVF attempt (back at that time that was quite successful considering that the first American IVF program in Norfolk went through over a hundred attempts before their first successful IVF pregnancy was born).

However, IVF is now a common method for the treatment of infertility that is an option for perhaps 15% of all couples experiencing infertility (an inability to conceive naturally). The last time I checked, there are over 250 IVF programs in the United States alone, and they are becoming more than just a group of physicians, embryologists, geneticists, and technicians treating infertility. As previously mentioned, preimplantation genetic diagnosis (PGD) is already being offered by many in vitro fertilization programs (IVF programs), and in the future, germ line genetic engineering will almost assuredly be applied through these programs. In other words, the clinical apparatus and infrastructure for genetic engineering are already in place. It is now a matter of deciding where we want to go with evolution. Evolution is literally now in our hands, and from a pragmatic perspective the genes involved, and the techniques will be developed in genetic laboratories, but they will probably be applied through IVF programs.

However, above and beyond the "hype" of genetic engineering or **Intelligent Directed (Design) Evolution**, what exactly are the genes we want to change?

Clearly there will not be a gene that produces anti-gravity and will make us fly. In addition, there are changes such as a young child with octopus-like arms that will not be in man's future as long as we have control of our own evolution. In Genesis 1:27, it is noted that man was created in the "image of God." Roughly translated, that means that mankind likes the way we look already…we look like a God (at least according to the Bible). I will simply ignore the arrogance of that assertion for the moment. However, the important implication of that biblical claim is that it will be doubtful that there are going to be radical changes in man's overall appearance. In other words, we like ourselves the way we are…at least in some idealized versions of ourselves like models, movie stars and/or some athletes or whatever "designer style" of Human that happens to be in vogue at the time. In fact, looking at some female models, even I have a hard time arguing with the concept that radical changes in Human morphology are probably not going to happen any time soon. In other words, multiple arms or two heads are not going to be in the game plan for the future evolution of mankind. Why? Simply put…we do not want it.

Then what will change about mankind with genetic engineering? The exact genes that will ultimately result in future mankind have not yet been delineated, but we can look at some broad categories, and in some cases give some specific examples of genes that will be on their way out or in, and ultimately look down the road to the new paradigm of Human evolution by genetic engineering.

Let's consider three broad categories of genes that will ultimately be genetically engineered into or out of the Human genome. I am going to use a simplistic and artificial "classification," but with the virtue that it will allow us to organize our discussion: 1) bad genes including copy number, insertions, or deletions, 2) SNPs (single nucleotide polymorphisms, variations on genes), 3) "designer genes," 4) junk genes, and 5) new genes.

1) Bad Genes

Bad genes are self-defining. They are genes that cause disease or are simply "bad" for mankind in one way or another. We can again divide "bad genes" into three subcategories: genetic inherited diseases, genes or alleles that predispose to a disease or that are associated with a disease, and less than optimal allelic variants. These three subcategories are really the same (they are all "bad genes"), but again, by dividing them in this way we will be able to consider some subtle but important differences, at least in the way or the sequence that genetic engineering will probably ultimately deal with each subcategory. In the future, with mankind in control of our own evolution, we will simply design these disease-causing genes out of the Human genome!

Genetic diseases are a rather obvious group of gene modifications that are easy to "predict," and will be genetically engineered in or out of the Human genome since efforts are already being made in this area, and with a few successes already reported.

There are a large number of genetic diseases such as cystic fibrosis, sickle cell anemia, hemophilia, etc. In most genetic diseases, there is a mutation(s) in the gene that results in a specific disease. There are thought to be about 21,000 genes in the Human genome, and a mutation that seriously affects the function of any one of the genes could potentially result in a "genetic disease" or perhaps death during embryonic or fetal development. These genetic diseases are in fact (generally) allelic (mutant) variations of normal genes, and usually result in some impaired or loss of function of the gene. Unfortunately, as we have previously noted, genetic diseases are evolution at work.

In fact, the problem of genetic diseases is even more complicated because we inherit two copies of each gene: one from our mother and one from our father. Sometimes disease only results if we get a bad gene from both parents (recessive transmission) or sometimes a disease results if we inherit only one bad gene (dominant transmission).
(http://www.nlm.nih.gov/medlineplus/cysticfibrosis.html)

For example, let's consider our "poster child" of genetic diseases. Cystic fibrosis occurs in one in 2500 to 3300 Caucasian babies in the U.S. (less in African Americans and Asians). It is the most common cause of chronic lung disease in children and young adults, and the most common fatal hereditary disorder in Caucasians in the U.S. Although there have been improvements in treatment, half of the people with cystic fibrosis are still dead by the age of 30. However, there are 1300 known genetic variations in the cystic fibrosis transmembrane conductance regulator gene (CFTR) that causes secretions to become thick and sticky and results in repeated infections and eventually lung failure…that is cystic fibrosis. In fact, we might say that there are 1300 different diseases known as cystic fibrosis, each associated with slightly different mutation and a slightly different function (or more precisely, a lack of function) of the CFTR gene. Both CFTR genes (one on each chromosome) must be mutated to have the disease (recessive transmission). If only one gene is mutated, that individual is a carrier, but will not have the disease. The carrier incidence in Caucasians in the U.S. is 1:25. If both the father and mother carry a mutated gene, then 25% of their children would be expected to have cystic fibrosis, and another 50% would be carriers like their parents, and 25% would be totally normal. The current PGD technique of "treatment" is to select an embryo for implantation that will not have the disease (ideally one of 25% normal embryos that is also not a carrier). Obviously, a better approach would be too "correct" the defective gene or insert a normal copy of the gene. In the case of cystic fibrosis, it may be possible to prevent the disease by simply inserting one copy of a normal CFTR gene. This might be done by inserting the gene on an artificial chromosome (discussed in more detail below).

Ultimately the "treatment" for cystic fibrosis will be to get all the defective CFTR genes out of the Human gene pool. However, the "cure" for all genetic diseases is going to be a tedious and long-term affair involving one gene at a time (at least initially, what that translates into is a few decades to sort out the genetics, but several hundred years or more may be necessary to realistically get rid of a genetic disease in the Human genome, since they will have to be attacked one gene at a time). In fact, the really tough problem is not going to be technology; the ultimate problem will be applying technology. We could completely eliminate cystic fibrosis from the world population right now with current technology, but that is realistically not going to happen. Eliminating cystic fibrosis would require a massive worldwide screening program to identify all carriers…think dollars. But then would come the impossible part; all carriers would have to go through preimplantation genetic diagnosis to ensure that the cystic fibrosis gene is not passed on…but again that scenario is highly unlikely to ever happen. In fact, this same scenario is true of many genetic diseases and even some infectious diseases.

For example, when I was a young physician there was a program to eliminate syphilis (at least in the U.S.). Syphilis is a sexually transmitted disease that can be easily screened for and easily treated. In addition, syphilis is a Human disease, and there are no natural reservoirs in the animal kingdom (unlike a disease such as influenza that can be found in many different animals, such as the bird flu, and periodically mutate back to a strain that can infect Humans). That means that if we find and treat all of the Human cases of syphilis, the disease will be gone forever. In addition, screening is rather easy since it is spread almost exclusively through sex, and most people know who they have had sex with. (The "joke" in medical school was the student who asked the question as to whether you could get syphilis from a toilet seat. The professor's answer was …yes, but you would have to be doing some strange and unusual things with the toilet seat.) Screening programs were set up, and whenever a positive test was found, that individual would be treated, but in addition, a government investigator would follow up and interview the individual's sexual partner(s) and trace (find) and test their contacts.

Positive contacts were treated, and again their contacts traced, and the process repeated. This was a simple strategy and technically very feasible. In fact, if applied internationally and aggressively pursued, such a simple approach could eliminate syphilis from the Earth. Did we eliminate syphilis? **NO!** Why not? Well…it's a sad story that boils down to money mixed with apathy and the "difficulty" in applying even a simple process worldwide. The Vietnam War started sucking up large sums of money, and few people really cared about syphilis. Besides, good church-going people did not get syphilis (despite the fact that I have treated many), and it was not a politically popular disease, and on and on. I treated a new case (not previously diagnosed or treated) of syphilis a short time ago. Syphilis is still with us…apathy prevailed. The technology was all there to eliminate syphilis, but the elimination of syphilis was not a technical problem. It was an applications problem. We could not consistently apply simple technology.

With that and other experience in mind, I have no illusions that we are going to eliminate the cystic fibrosis gene or any other genetic disease (or syphilis) any time soon. Apathy will prevail. We will simply continue to chip away at a large mountain with a hand chisel…one chip (case) at a time. Again, the problem is not science or technology; it is the application of science and technology. If the future approach to genetic engineering is to "treat" one genetic disease at a time, then genetic engineering will have little or no impact on the evolution of mankind for many years or perhaps never. It would be (at best) just another medical technique…like performing an appendectomy. It would be both good and important to the individual "treated," but it would translate into very little in terms of evolving mankind. However, there are alternative approaches, which we will discuss below, but they will require an entirely new paradigm.

Preventing genetic diseases is currently actively being pursued, and those will undoubtedly be the first genes in the germ cell to be altered by genetic engineering. In fact, that is exactly what PGD is all about, however, we should recognize that PGD is very primitive, treats only one problem (gene) or at most a few diseases at a time, and it is going to be a costly form of "genetic engineering." In fact, in some respects, I am reluctant to even call PGD a form of genetic engineering. Currently this is a technical problem, but once the techniques are worked out, it will turn into an applications problem, which, as noted with syphilis, we (as a Human population) do poorly, and it could take hundreds of years (if ever) to have any significant impact on the Human gene pool or anything that we could call "evolving mankind." Eliminating one genetic disease from one individual is medical treatment, not evolution. (http://www.ncbi.nlm.nih.gov/books/bv.fcgi?call=bv.View..ShowTOC&rid=gnd.TOC&depth=2)

There are a lot of genes that will have to be eventually "engineered," not just the genes known to be associated with hereditary diseases, but the genes that predispose to cancer, atherosclerosis (one cause of heart attacks), Alzheimer's, etc. If we alter all these genes, we may not look any different, but we are going to be much healthier and live longer, simply because we will not be subjected to a wide variety of diseases. Since almost every disease has a genetic component, this group will ultimately turn out to be a very large group of genetic variants. However, one reason that I have separated them from the hereditary diseases is that we really do not know at the present time what genes are associated with which diseases, and as I have previously noted, it is going to take a long time to figure this puzzle out. Discovering the relationship between all the diseases mankind is subject to and their genetic relationship may take a larger part of the 21st century. In fact, I would predict that the next hundred years of medicine will be heavily involved with this exact problem of determining which allelic variants

are involved with which diseases, particularly with the disease of atherosclerosis (heart attacks, strokes, and peripheral vascular disease), which is associated with 50% of current Human deaths (at least in the U.S.), and cancer (again, associated with roughly 25% of (U.S.) deaths. However, this "subcategory" of genes has the potential to at least equal the number of genetic diseases and quite probably even surpass them in number. The problem is that we have only recently begun to recognize the fact that most of the diseases that the Human population suffers from are in fact related to our genetics in one form or another. Medicine has made great strides…especially against infectious diseases (other organisms [generally microscopic] trying to eat us for dinner), but also great strides in understanding the anatomy, physiology and molecular biology imbedded in the genetics. However, we are only beginning to understand the genetics behind many diseases, even very common diseases like atherosclerosis. The next advancement in medicine and science must be in understanding genetics, and the relationship of different genes to disease and health, intelligence, and longevity.
(http://www.motherjones.com/news/special_reports/1998/05/marshall.html)

We cannot do genetic engineering to prevent diseases until it is clear which genes or allelic variations are associated with which diseases, and that is what is currently being worked out. This will turn out to be the limiting factor on any rapid or widespread application of genetic engineering for the prevention or treatment of diseases, and anything close to what might be called evolution (**IDE$_H$**). Multiple studies to relate genes or allelic variations to diseases are really just beginning. For example, the famous Framingham (Massachusetts) heart study that has been ongoing for 58 years is planning a study scanning for 500,000 single nucleotide polymorphisms in 9,000 study patients, looking for genetic variations predisposing to disease (or lack of disease). The National Cancer Institute has announced a whole-genome scan for prostate and breast cancer genes, and the Department of Health and Human Services had proposed 68 million dollars in its 2007 budget to study genes and the environment. (Kaiser,

J. *Science*, vol. 311, February 17, 2006.) One private company (Decode) has obtained the exclusive rights through the Icelandic government to the genes of the people of Iceland. Iceland's 270,000 people are a very homogeneous population with relatively few outside influences for hundreds of years, and it will be easier to identify genes associated with particular diseases. In addition, Iceland's genealogical records are computerized and in some instances date back a thousand years. (Marshall, E. *Science,* May/June 1998.) Decode plans to identify the function of genes that contribute to 45 common diseases. Decode has already identified variations in the gene encoding a 5-lipoxygenase activating protein (FLAP) that doubles the risk of heart attack. Likewise, GenBank, which is a branch of the National Library of Medicine (which is a branch of the National Institute of Health), has also identified several different alleles of the apolipoprotein E gene (ApoE) that carries cholesterol in the blood, and that predisposes to atherosclerosis.

In addition, allelic variants of two genes are thought to account for three-quarters of all cases of age-related macular degeneration, which is the leading cause of blindness in people over age 60. Macular degeneration affects over 60 million people worldwide. (*Science News*, vol. 169, no. 10, March 11, 2006.) Rando Allikmets of Columbia University has reported that certain alleles of a gene for a protein called factor H increase the risk of age-related macular degeneration, while other alleles protect from the disease. Factor H is involved in shutting down inflammation. In addition, another gene called factor B was identified that turns on the same pathways that factor H turns off. There are also gene variants (alleles) of factor B that increase and decrease the risk of macular degeneration.

These examples are really only the beginning of the data that is going to be necessary for literally every gene and every allelic variant. We are literally going to have to rewrite the medical textbooks to include all of the genetic allelic variants and their relationship to all of the known diseases. But then comes the hard part...getting those disease-associated alleles out of the Human genome.

2) Single Nucleotide Polymorphisms (SNPs)

Alleles are genetic variants of a gene. However, almost every gene has multiple slightly different (mutated) forms spread out amongst the Human population. A SNP is a Single Nucleotide Polymorphism or variation of a single nucleotide at a particular location in a gene. They are literally small variations that may or may not affect a gene's function. This is truly an evolution in progress. Most of the genes that we have already discussed as "genetic diseases" or genes that predispose to diseases are simply alleles of normal genes. However, within the Human genome, as we have noted, it is estimated that there are ten million single nucleotide polymorphisms (SNPs). In other words, you might say that there are really ten million Human genetic variants. In fact, since each of us inherit some combination of all the allelic variants, we are all genetically unique (except for identical twins). However, not all these SNPs are in "genes." Here, I am using the term gene for transcribed and translated DNA. Many of these polymorphisms are in junk DNA. What is just starting to be worked out now is how all these gene variants are associated with both genetic diseases (considered above, but which are really only allelic variants that result in clinical conditions that we call a disease), which alleles predispose to diseases, which alleles are less than optimal, and which ones are in essence truly "neutral" or optimal. In fact, that is exactly what the "less than optimal" allelic variants are all about. There will be thousands of allelic variants found that are not exactly disease-associated but are also not the ideal allelic variant that we want in the Human gene pool...they are "less than optimal" variants. Unfortunately, like the disease-associated genes, it will take a good portion of the 21st and probably the 22nd century to discover exactly which allelic variants are "less than optimal," which ones are neutral and which ones are optimal. In fact, it will probably be more difficult to sort all the variants out simply because they are not associated with an obvious disease.

They are an example of at least one form of evolution in progress. However, many SNPs result in poorer function of a gene, which may result in a predisposition to a disease or even death. There is now a bioinformatics web site called SNPedia that has cataloged 110,000-plus different SNPs (as of April 2019) with information of reported medical conditions. However, there are an estimated ten million SNPs in the Human genome, so we still have a long way to go before understanding the full ramifications of SNPs.

Understanding and characterizing SNPs allows the development of personalized medicine. For example, there are 100 million prescriptions a year for a class of drugs called statins used to reduce blood cholesterol. However, some people develop a myopathy (muscle inflammation) while taking statins that can sometimes even result in death. Dying from taking a drug to improve one's health and decrease the risk of a heart attack or stroke is problematic. However, a single SNP has been identified that increases the risk of myopathy by 5 to 17 times. This will allow doctors to test for this allelic variant before prescribing a station. However, cost may limit the application of this advancement in diagnosis. This field is developing very rapidly and is associating genetic variants with a predisposition to many diseases and will allow a more personalized medicine. Individuals will know what diseases they are prone to in advance and may be able to take measures to limit or avoid health problems. However, when we understand various SNPs, it may be possible to genetically change SNPs associated with disease, and in addition, allow these genes to be deleted out of the Human genome. I wish to emphasize that this is not some future development. It is going on right now. What the future will certainly bring will be a much better understanding of the involvement of SNPs in disease and health, and ultimately attempts to eliminate "bad" SNPs from our children, and ourselves using genetic engineering.

3) Designer Genes

The genes that make us look the way we do are often called "designer genes." Designer genes are the genes that opponents of genetic engineering love to hate. Why? Because basically these are the genes that the usual opponents to genetic engineering drag up as useless and irrelevant results of genetic engineering, and of course they are absolutely right within their "straw man" context. I agree that the "choice" of things like eye color, hair color or skin color are as trivial as the latest clothing fashion (and the "choices" of designer genes will probably change over time just as clothing fashions change), and if that was all genetic engineering was about, it would be a worthless and useless undertaking. This is where the label of "designer babies" comes into play, which is also the source of a great deal of criticism of genetic engineering.

However, the positive side of designer genes is that they will "level the playing field of life" …literally. Even though we like the way Humans look, in general, we do not all look like movie stars, or professional athletes, or swimsuit models. In other words, we do not all look ideal or perfect. In fact, if I were to be blatantly honest to the point of rudeness…most of us do not look anything like perfect…even those in Hollywood. Moreover, the composite effect of "designer babies" may not be so trivial. Most of the world's female population does not look like a Bo Derek "Ten," and there are very few men in the world that could walk onto a basketball court and seriously challenge Michael Jordan, even if they practiced basketball for an eternity. Most of the world's population does not have the right set of designer genes. This takes nothing away from Michael Jordan who has been called the greatest basketball player…ever. He did not get where he is by "sitting on his genes," so to say. However, it does recognize that many of our "stars" in every field of Human endeavor often have a "genetic advantage" in one form or another…good looks, athletic ability, physical build, intelligence, etc. So, should we genetically engineer every new baby to be a Michael Jordan and a star basketball player, or every woman to be a model or movie star? Of course, not… and it will never happen. However, there is also nothing wrong with making our children look more like the "Gods" that we think that we are supposed to look like. This will result in a "leveling of the playing field of life" giving every newborn baby the best set of "designer" genes. Look in a mirror; if you think that your child should inherit all of your defective genes without any improvement, then you are either one hell of a person (God comes to mind), blind as a C. elegans (no eyes), or dumb as a sponge (no nerve cells), or…perhaps you hate your children. "Designer genes" is nothing more than giving our children the best genes possible and to "level the playing field of life." Yes…I agree as noted above that if we simply look at some of the details like hair color, eye color, skin color, etc., such changes would be trivial, and not something worth changing the genetics. However, when the techniques of genetic engineering are perfected and are safe and reliable, there is nothing wrong with giving our children the best set of genes possible, including a particular set of designer genes. Selecting a group of designer genes will not make anyone a movie star or a model or an all-star basketball player, but the right genes would make any of those things possible. I can absolutely guarantee you that if Michael Jordan had my genes rather than his, he would not be the world's greatest basketball player…wrong genes for the job. So, Michael…don't worry, I will not be taking your job away from you anytime soon.

Let me give one concrete example of a designer gene that I think is almost guaranteed to be eventually engineered into (out of) the Human genome. It is the *Myostatin* gene, which is responsible for maximum muscular development. This is not exactly my favorite example of where the Human race should be going genetically, but considering athletic competitiveness and the number of people who spend years working out in gyms and lifting weights trying to "bulk up," there is a high probability that knocking out the *Myostatin* gene will be high on the designer gene list (provided that we do not discover some disaster resulting from the *Myostatin* gene expression in the future.) Whenever we exercise, jog long distances, play sports or do any physical activity more vigorously or of longer duration than we are used to doing, muscle tissue is "damaged." Muscle fibers have multiple nuclei. When muscles are stimulated by exercise, satellite cells around the muscle proliferate and fuse with the muscle fiber. When you feel sore for several days after strenuous exercise, your muscles have been "stressed," and are undergoing hypertrophy (they are getting bigger). You have probably heard the expression: "No pain, no gain." After hypertrophy, the muscle fiber is bulkier (and stronger). A protein called insulin-like growth factor one (IGF-1) stimulates the proliferation of the satellite cells, which helps the muscle get bigger (hypertrophy). Interestingly, *Myostatin* has the opposite effect and limits hypertrophy. You probably did not realize that when you exercise to increase muscle mass, you are literally fighting against the *Myostatin* gene that is trying to prevent you from "bulking up." Some might think that it is rather dumb of Mother Nature to invent such a useless gene...think again. When you have twice the muscles that you need, you have to feed that extra muscle mass. In a world where starvation has always been just around the corner, the *Myostatin* gene is a very smart gene to have for survival. This is probably why the *Myostatin* gene is widely distributed in the animal kingdom from mice to men. Those creatures that did not have it are no longer with us...their ancestors starved to death long ago (perhaps with lots of muscle) because they had to feed more muscle bulk than they needed for survival.

However, in our current society, the *Myostatin* gene has become something of a liability to those who want to increase muscle mass. Most of us are not on the brink of starvation. Would blocking the *Myostatin* gene by genetic engineering be a desirable "designer gene" change? The answer to that question would undoubtedly depend on who you ask. If we ask professional athletes or people who spend a significant amount of time trying to "bulk-up," we would probably get a lot of favorable responses to blocking the expression of the *Myostatin* gene. In fact, there are attempts to find a drug to block the *Myostatin* gene right now. The Belgium Blue bull shown below has a natural non-functional mutation of the *Myostatin* gene. Would we be leveling the playing field of life if we all had the anatomy of Arnold Schwarzenegger without ever setting foot in a gymnasium?

THE FUTURE OF HUMANITY IS NOW

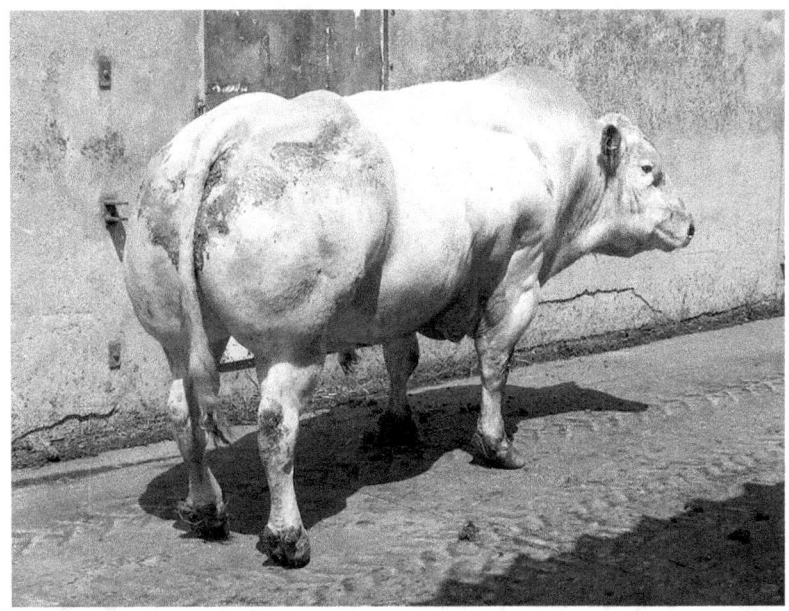

The Belgian Blue is a type of cattle known to have a non-functional mutation of the *Myostatin* gene, which results in massive muscles.

Image Credit: Wikimedia Commons, Public Domain
(https://commons.wikimedia.org/wiki/File:GEURTZ16.JPG)

Interestingly, we do not have to do any experiments to see what such a change would be like in Humans, nor do we have to discuss the "moral rationale" for some scientists to produce a child with a loss of function mutation of the *Myostatin* gene. It has already been done in Humans…naturally. In 2004, there was a report of a boy born in Germany who was excessively muscular even as a baby. He was found to have a loss-of-function mutation of both alleles of the *Myostatin* gene (like the Belgium Blue bull above). Mother Nature has already done the experiment. However, the long-term effects are not known, and the boy is being closely followed by physicians.

When looked at in the broader perspective, I do not think that we have to apologize for genetically engineering designer genes into a baby. It is only a matter of "leveling the playing field" of life, and making all of us look more like Gods, and not just a few of us...all of us. The genetic engineering of designer genes will not make Humans any "better," they will simply make us look better and give us the genetic "edge" for many Human endeavors such as athletics or perhaps make us look more like Miss America.

4) Junk Genes

In addition, re-engineering the whole Human genome will probably involve eliminating what has been called "junk DNA." The Human genome has been characterized as composed of over 90% "junk DNA." However, many previously considered "junk genes" are now being recognized as control genes (siRNA and many other transcribed, but not translated into DNA) or DNA with other functions. Although this will probably not be a high priority "fix" for genetic engineering, since they do not (at this time) appear to cause any known diseases, most junk DNA will probably eventually be removed. We previously mentioned the fact that there are about 850,000 LINE elements in the Human genome and many other types of DNA fragments (junk) that act like vertically transmitted endogenous viruses (we are not infected like influenza, we carry the genes in our genome). Unlike a virus that has to attach to a cell, get into the cell, shed its "coat" and replicate itself, and then reassemble multiple copies of itself, and get back out of the cell, and then find a new cell to repeat the infectious process again, these "vertically transmitted" virus-like DNA pieces never have to leave the comfort of their home in the Human genome (or in the genome of many other lifeforms from mice to rice). Yet, every newborn baby has a full new set (vertical transmission), and every cell in the baby as it grows also gets a full set of copies. They may (or may not) be hurting us, but they are a considerable metabolic and energy drain, and...who needs them? Bacteria have very little junk probably because in the competitive environment in which they live; wasting energy replicating DNA that is not necessary for their survival would put them at a significant disadvantage to other bacteria that do not have junk DNA. That is another way of saying that those bacteria that once had significant amounts of junk DNA or began to collect junk DNA are no longer with us. On the other hand, rice and corn have huge amounts of junk DNA, and have managed to survive.

The only positive role that has been suggested for junk DNA is that if a virus infects a cell and inserts its DNA into the cell's DNA, the probability is that it will be insert into the junk DNA rather than into a gene coding for a vital cell function, simply because there is much more junk than genes that are transcribed and translated. However, if this is the only thing junk DNA does for us, we will probably ultimately want to get the junk out of the Human genome and make our genome a leaner and more efficient genome. Let these freeloaders make a living somewhere else. The other possible role for junk DNA which we have previously mentioned is to remove reactive oxygen radicals and other mutants. Each cell produces many oxygen radicals that can react with DNA, causing mutations. Junk DNA may serve as a pool of DNA to neutralize the oxidative radicals that cause mutations. However, removing all the junk may make it possible to cut the Human genome dramatically in size, and in addition add genes that can increase many human functions. This could become an important consideration when we get to the probable ultimate paradigm of genetic engineering (discussed shortly). We could add more genes that are specifically to reduce or neutralize oxygen free radicals, viruses, and other mutants. This will have to be looked at carefully. We will be discussing this much more later.

5) New Genes

There is one other broad category of possible genes that will be genetically engineered into the Human genome, and those involve changes in genes that will alter mankind in ways that we do not currently have examples that we can point too because they do not currently exist in the Human population. We might call these "new genes" even though some could be simply modifications or expression of existing genes or genes taken from other species. Probably the two most important in this category will be genes to markedly increase Human intelligence, and genetic changes to prevent or at least markedly alter the aging process (in addition to specific genetic changes to cure diseases).

Although the prospect of increasing Human intelligence may appear to be science fiction, this prospect may turn out to be the easiest to accomplish and the most important genetic engineering modification that will literally produce a "new species" of man. For example, Joe Tsien from Princeton University has genetically engineered a mouse that is 4 to 5 times as smart (at least in terms of memory) as a normal mouse by inserting a single gene into the mouse genome. What about brain size? If we increased the size of the cerebral cortex, could we increase our intelligence? Although there are problems with just considering brain size and equating that to intelligence (which we will discuss later), there is no question that the last several million years of Human development has been associated with an increase in brain size, particularly of the cerebral cortex. However, the question remains: can we realistically genetically engineer an increase in brain size? It has already been done. Chenn and Walsh generated transgenic mice to over-express a modified protein by neuroepithelial precursor cells in the brain. (Chenn, A., Walsh, C.A. "Regulation of Cerebral Cortical Size by Control of Cell Cycle Exit in Neural Precursors." *Science*, vol. 297, July 19, 2002.)

The results were a marked increase in the mouse brain size…particularly the cerebral cortex. We will look more closely at some of these possibilities when we discuss the future of Human intelligence in the chapter *The Wise Monkey*. I would opine that increasing mankind's intelligence will be the most important evolutionary advancement that can be accomplished by genetic engineering…even more important than curing diseases and extending our lifespan. However, the human genome is mainly junk DNA much of which has gotten into the human genome a very long time ago. This is also true of animals such as mice and rats and almost all animals. We will discuss this much more later.

The prospect of using genetic engineering to significantly extend the lifespan of mankind will also be discussed in more detail in the chapter *Methuselah* but will probably include genes to better repair oxidative damage and more faithfully replicate DNA, genes that produce better repair of DNA damage, extend telomeres, repair or replace damaged mitochondria, and altering or modifying genes that are associated with the aging process. In fact, there is probably no "single fix" or single gene for aging, and any significant extension of lifespan will probably require alteration or insertion of dozens or perhaps even hundreds of genes. There is nothing "in principle" that should prevent an animal from living indefinitely. Living indefinitely will probably be a realistic goal of the future. Unfortunately, religious groups have gotten involved in many of these areas and have made it illegal to modify the genome of children. They have apparently managed to get something similar passed supposedly in China.

However, exactly which genes will be required to slow of modify aging is only now beginning to be worked out, and we do not know if we can increase intelligence above an IQ of about 200.

In addition, there are many interesting genes that may be found in other creatures, from bacteria to other vertebrates that may be beneficial to mankind. For example, some bacteria have enzymatic pathways to eliminate (break down) environmental toxins or better repair damaged DNA that might ultimately be beneficial for man. The rat has an awesome immune system to fight infections, and the salamander can regenerate a leg that has been amputated. Some of these genes might benefit mankind.

Creating an entirely new gene to perform some useful function and control its function sounds like a "great idea," but the complexity of creating a new gene for a particular function without some similar gene already present in some creature's gene pool to use as a model will be a considerable genetic feat. This group of potential genes will probably be the last to be utilized. However, there have already been attempts to do exactly this.

Another important gene is the TP53 gene 0n the short arm of Chromosome 17. The various proteins isoforms produced by the TP53 gene are called P53 protein and are tumor suppressors. However, about 50% of the P53 isoforms are mutated in cancers. For example, the cancer mortality in elephants is about 4.81 percent despite their considerable mass compared to 11 to 25 percent cancer mortality in humans. However, humans have one copy (2 alleles) of TP53 compared to 20 copies of TP53 (40 alleles) in the elephant. Therefore, despite the very large difference in the number of cells in the elephant that could become cancerous compared to Humans, the caner incidence is much lower. This difference in the incidence of cancer is thought to be due to the much larger incidence of the TP53 genes in the Elephant.

Genetic engineering in these seven areas has the potential to remake mankind. They are not all "pure" categories. Some categories tend to blend into one another. For example, treating a "genetic disease" is not much different than selecting a particular gene allele or set of alleles that are associated with a very low risk of cancer or atherosclerosis. Likewise, (to be discussed shortly), the first real increase in intelligence may result not form insertion of a particular new gene for intelligence, but from selecting of beneficial genetic alleles for brain size and intelligence already present in the Human gene pool. In other words, we give the brain a new structure, and hopefully it will utilize the new structure to increase intelligence.

Unfortunately, there are many that would prefer to have mankind remain the puny, pitiful, and pathetic species that we are (by comparison to what we can realistically become using **IDE**$_{\text{H}}$), and they will unfortunately (for them and their descendants) have to be left behind just as the Neanderthals were left behind 25,000 years ago. This is another area where we can apply the refrain of the gospel of Matthew: "For many are called, but few are chosen." (Bible, KJV, Matthew 22:14)

Genetic Engineering is Here Now!

Although I have tried to emphasize the importance of knowing what to genetically engineer, as well as developing effective techniques to alter, replace or "knockout" genes. Genetic engineering in bacteria, yeast and single-celled eukaryotes has been around for some time. However, there have also been multiple attempts to genetic engineer genes in Humans. These attempts have been limited to somatic cells rather than germline cells (egg, sperm, or early embryos). Unfortunately, the results have not been very good. Although it is obvious to almost everyone that the potential for treating genetic diseases and many, if not most, other diseases is considerable, the initial reality has been disappointing...in fact the word disastrous comes to mind.

On September 17, 1999, eighteen-year-old Jesse Gelsinger died as the result of an attempt at gene therapy. Jesse was suffering from a rare genetic disease called ornithine transcarbamylase deficiency (OTC). Half of babies born with OTC die in the first month of life. An adenovirus (a virus that causes the common cold) has been engineered to contain the corrective OTC gene. Viruses have been frequently used as vectors to carry a desired gene into cells. This approach seemed to make sense, because viruses have "evolved" over millions of years to do exactly the job of carrying their DNA into cells. Unfortunately, in Jesse's case, he developed a severe immune response to the virus, resulting in multiple-organ system failure, and he died four days after receiving the injection of the adenovirus. Jesse's death put a hold on gene therapy and raised the issue of what type of "vector" can be used to safely carry a gene into cells.

Another gene therapy trial involved children with a severe form of immunodeficiency...the so-called "bubble boys" (X-linked SCID), who have to live in a sterile environment, or they will die of infections. Nineteen infants were treated with restoration of their immune system, but five of them developed leukemia, and one died. This is thought to have resulted from the random insertion of the therapy DNA into cellular genes, which can result in non-functioning of the cellular gene in which the insertion occurred and can result in loss of function of cancer suppressor genes. This raises the problem of random insertion of genetic material rather than targeted insertion at a particular site.

However, successes have begun to be reported. A brain disorder in boys due to a genetic lack of a protein that helps maintain the myelin sheaths around nerves has been successfully treated in France by inserting a corrective gene into two boys. The progressive brain damage had apparently stopped during a two-year-period following the gene therapy. (*Science*, vol. 326, 18 Dec. 2009.)

In addition, twelve patients with a rare form of inherited blindness have shown improvement after injecting a gene needed to make a light-sensing pigment. The gene was enclosed in a virus and injected directly into the eye. Four children gained enough vision to play sports and stop using learning aids at school.

However, the subject of genetic engineering or what is starting to be called gene editing has very recently taken a dramatic step forward!

These successes confirm the potential of gene therapy to be a powerful tool for correcting many diseases. However, it is also clear that vectors used to insert correcting genes and the technique of random gene insertion currently leaves a lot to be desired. Gene therapy will have to become routine, and well understood, before any serious attempts at **IDE**$_{n}$ can be considered. However, it is also clear that the current paradigm of Human evolution is and has been a total disaster resulting in most of the diseases that Humanity currently endures including aging, a short lifespan, and less than an optimal intelligence level that has resulted in the majority of the world's population living in a mythology where hunger, poverty and war abounds, and even with the pathetic destruction of the environment of the Earth itself. We must remake a healthier and much smarter Human…time is running out.

A New Paradigm

From the preliminary data that is beginning to be published regarding "bad genes" or allelic variants associated with disease, it is starting to become clear that almost every new baby is probably inheriting thousands of disease-associated alleles. This can (and probably will) be initially or ultimately viewed as a medical problem, and eventually attempts will be made to "cure" some of the resulting "diseases" using genetic engineering to eliminate or modify such genes. However, even when all or most of the technical problems of genetic engineering are solved, the approach of trying to treat each of the hundreds or thousands of these "bad alleles" will turn into the "applications problem," and as we have noted, will probably take hundreds of years to improve. In fact, as previously noted, this will require medical genetic engineering, which would be important for the individual, but will have little consequences for the evolution of the Human race until and unless it is continued for long periods of time. In addition, trying to change every disease-associated allele is almost guaranteed to grow into a problem that will be unmanageable. In other words, genetic engineering is probably not going to be a simple or "quick fix."

The idea is to identify all the genes that result in genetic diseases and all of the other diseases and also all of the less-than-optimal alleles and eliminate all of them from the Human genomic pool not just one at a time, but all at once. In fact, it may be technically a lot easier to eventually synthesize a whole new Human genome using what we have called the **"Optimal Human Allelic Set(s)" (OHAS)** of genes and copy it millions of times to use as "stock" rather than try to alter thousands of different disease-associated alleles for each new baby. The only thing that would have to be altered in the **OHAS** would be the "designer genes" for hair and eye color, body build, height, etc. chosen by the parents. In addition, we would need to add any "new" or transHuman genes that are discovered, and unfortunately, that will have to be done one baby at a time, but not necessarily one disease at a time. In fact, if we consider all of the genetic diseases and genes that predispose to a disease that would have to be modified and all of the less-than-optimal allelic variations that we would need to change, and adding new genes to increase intelligence and longevity, and possibly even getting rid of the junk DNA that has accumulated in the Human genome, it becomes obvious that the ultimate approach to genetic engineering will be to synthesize the Human genome de novo. Although this sounds like a formidable task (and it will be), it may ultimately be simpler than trying to change hundreds of genes in each egg (oocyte) cell and sperm or each fertilized one-cell embryo. The "bottom line" for genetic engineering will be a set of "ideal" gene alleles that do not cause genetic disease, and are not associated with any disease process, but are the **Optimal Human Allelic Set(s)** of ideal genes. However, we must recognize that the **Optimal Human Allelic Set(s)** without "new" genes to increase intelligence or provide a longer lifespan will not make us more than we are now. Each Human would be optimized of sorts, but still a man (or woman).

Of course, we cannot technically do any of this at the present time. We have not even identified the functions of all the genes in the Human genome, let alone know which ones are associated with which diseases or which allelic variants are ideal (discussed above). There is a reasonable probability that it will take most of this century to work out the details of what will constitute the **Optimal Human Allelic Set(s)** of genes, and ultimately the new genes that will markedly increase man's intelligence and increase lifespan and prevent aging and eliminate many if not most diseases. In fact, it is almost guaranteed that there will not be just one **Optimal Human Allelic Set** but many variants. In addition, as we will discuss shortly, the human genome is not an incredible collection of genes that make us the smartest creature on Earth. In fact, as we will discuss shortly, the Human genome is full of garbage.

An Absurd Pipe Dream

Absurd... not at all. When I was young, there was this crazy idea of a man going to the Moon. That was absurd! It was of course only science fiction. Going to the Moon was fantasy, not science. There were "canals" discovered on Mars (which we now know do not exist since we now have rovers on the surface of Mars). The Wright brothers ran a bicycle shop in Dayton, Ohio (I lived not far from their house). They introduced flight in 1903 in an "aircraft" that looked like a bunch of egg crates held together by a tangle of wires. When I was young, I would rather have ridden one of the Wright brothers bicycles rather than have considered going to the Moon. However, things changed. Soon planes were flying everywhere and going to the Moon changed from a dream into a reality.

Robert Goddard began experimenting with small rockets in Massachusetts and launched the first liquid-fuel rocket in 1926, and Hitler unfortunately managed to reshape London's architecture with German rockets in the 1940s. On July 20, 1969, Neil Armstrong and Edwin Aldrin walked on the Moon. In 66 years after the Wright brothers first flight, men walked on the surface of the Moon, and returned safely to the Earth...that was absurd!

Why did it happen? Because man wanted it to happen, and we rapidly developed the knowledge and engineering capability to make it happen. The desire to land a man on the Moon and the engineering capability converged in the late 1960s. Even though to a realistic man not prone to bouts of fantasy (someone with his/her feet obviously planted on the ground), going to the Moon was clearly an absurdity in the 1940s and 50s, and an even greater fantasy before the advent of flight in 1903. However, there was nothing **"in principle"** that prevented man from going to the Moon. The restraints were very difficult technical problems as seen from the perspective before the 1960s (or even during the 1960s for that matter), and in addition, no one really believed it was possible except perhaps a few science fiction writers and some physicists. The problems were complex and involved technologies that were beyond our capabilities at that time and involved rocket engineering, telecommunications, spacesuits, "G" forces, guidance, rocket exit velocity, and the little problems like breathing on the Moon, and the physics of re-entry with temperatures that would literally melt known or then conceived spacecrafts. We tend to forget how difficult those problems were with the knowledge, and scientific and engineering capability available at the time. For example, Ray Kurzweil (and others) have pointed out that there is much more computational capability in the average modern automobile now than on the Apollo spacecraft that went to the Moon. (Chou, Stephen. *Technology Review*, Feb. 2002, p 41.)

How times have changed, yet we sometimes forget how far we have come. **What man desires, and what can be done "in principle" within the laws of physics is potentially now within our grasp.** Genetic engineering is still at a very early and primitive state, but we have begun to fly, and we have gone to the Moon, and we are on the verge of re-engineering ourselves. There have been radical advances in science and engineering. You can count on a lot of changes in the future that sound like science fiction now.

Why Does Mankind Need To Evolve?

Today, at least some of the infrastructure for **Intelligent Directed Evolution** is already in place, but it resembles the first Wright brothers' plane. However, there is nothing "in principle" that will prevent man from reshaping man and begin to intelligently direct the evolution of mankind in the future. But do we really need improvement? After all, we arrogantly believe that man was created in the image of God as noted in Genesis 1:27; *"So God created man in his own image, in the image of God created he him; male and female created he them."*

Is **Intelligent Directed Evolution** a desirable goal? Each of us will need to answer that question for ourselves. I would simply point out that as the most intelligent creature on the Earth, we have literally become the stewards of this planet, and also stewards of the other creatures of Earth…including ourselves. We have fortunately or unfortunately evolved ourselves into the creatures responsible for all life on the planet Earth and even the Earth itself—our environment. Yet, mankind does not appear to have yet evolved enough to handle all of the problems that appear on the horizon.

In your honest opinion, has man as a **work in progress** (which we have roughly defined as man evolved to the present time) done a good job in relation to our environment and the other creatures of this planet? What about our fellow man?

Have we, Homo sapiens sapiens, followed even the most basic concepts that Jesus Christ (and many other great thinkers) proposed thousand years ago with the Golden Rule—love thy neighbor as thyself. Has mankind followed the teachings of the Buddha and the way of the Eightfold Path of Buddhism? Let us even forget about Gods. Even a cursory look around today and especially any consideration of history gives a clear answer... **No!** Instead of following even simple principles proposed by Jesus, Buddha or by dozens of others great social thinkers, mankind has spent our time killing people who did not believe in Jesus or Buddha or some other God or some other politics. We have spent the ages killing everything around us, including ourselves, and we are a plague upon our environment and our planet. Mankind, at the present time, is the scourge of the Earth. How could man go so wrong? I lay the blame clearly on the current limited intelligence of mankind. We have not been smart enough in the past to handle the job of being stewards of the Earth. We have not even been smart enough to develop the social intelligence to control ourselves. Wars over a multitude of political and religious trivia have killed millions of Humans. Yet, we have the job. Man is mankind's worst enemy, and mankind's biggest threat. Mankind is well on its way to ruining the Earth's environment, and we have decimated many of the other creatures that have had the misfortune to share the planet with us.

Consider the inquisition in which many (particularly Jews) were burned at the stake. Many women were killed as witches between 1275 B.C.E. and as late as 1894. I have seen estimates of up to 9 million people killed as witches, but I am reluctant to accept these numbers, since they are at best guesses, and no one seems to have any good data. However, religious advocates often dismiss the inquisition or killing of witches or even the crusades as grossly overexaggerated and involving only a few "incidences." I mention these examples because they are killing in the name of a God—often Jesus Christ. But how can so many "believers" kill in the name of Jesus Christ who (we are told) specifically preached a philosophy of "love thy neighbor" and "turn the other cheek," and the Golden Rule: do unto others as you would have them do unto you. The philosophy and social principles behind religion and particularly Christianity are great, but the application has often been catastrophic, and reflects a limited "social intelligence" (to be discussed again in 3-5 *The Wise Monkey*), and clearly a "flaw" in basic Human logical thinking. Not one but millions of Humans have been able to accept as a guiding principle the concepts and values taught by Jesus of Nazareth, yet behave and function in a manor completely contradictory, and never "see" the conflict in logic—yes, logic.

Note that even though I am using Christians as the example here, the same concept could be applied to many other beliefs—religious, political, or otherwise. There have been many "good ideas" for Human social behavior proposed. I should also point out that what we are discussing has nothing to do with Jesus or his teachings. Jesus is clearly not responsible for the atrocities committed in his name. What we are discussing is how advanced (or limited) is the intelligence (or more precisely, the lack thereof) that can hold a belief— "thou shalt not kill," and go off to war to kill in the name of God or some difference in political or other trivia. This, again, is a variant of doublethink—the ability to hold two separate and contradictory ideas at the same time without recognizing a conflict. This defect in intelligence (my characterization) was recognized and popularized by George Orwell in his book *1984*. In other words, current Human "intelligence" allows most of the world's population to believe totally conflicting concepts (or hold ridiculous ideas or belief as true, and yet act completely contrary to the principals involved) without noticing that their beliefs and actions are logically inconsistent and incompatible.

Even a "ballpark" estimate of the number of Humans massacred by other Humans is almost certainly in the hundreds of millions, if we include wars—tribal, racial, national, religious, and interpersonal disputes. I place the blame for this pogrom on Humanity by Humanity on a lack of higher intelligence, and especially a lack of "higher" social intelligence.

With the increased intelligence that mankind has evolved in the past six million years (since we separated from our last common ancestor with the great apes or chimpanzees) has come knowledge and power. We have literally eaten from the tree of the knowledge of good and evil. In other words, we have developed a moral component to our intelligence. But with the knowledge of good and evil has also come responsibility. Yet, Homo sapien sapien has clearly not demonstrated a high enough level of intelligence to meet the challenges presented to us. Man is king of the intelligence pile on Earth in our present time epoch…yes, but it is a very small pile. We clearly need more intelligence, a lot more intelligence, and we clearly cannot wait another million years for more intelligence to evolve. Mankind desperately needs to re-make man especially by increasing our intelligence, but also in other ways to fulfill our basic ethical responsibility to ourselves, to other creatures on Earth and to the Earth itself. **The remaking of man by man is not just a desirable goal; it is an absolute necessity for our survival, and the survival of other creatures on Earth and even the Earth itself.** Forget about designer genes, we need to remake mankind for pure survival reasons, and to fulfill our responsibility to ourselves, the other creatures on this planet and to the Earth itself.

When we view intelligence as the ability to reason deductively, solve problems, think logically—literally the ability to take some data and draw some logical conclusions—it becomes apparent that many Humans (and there are unfortunately a considerable number) have an impairment in intelligence because they are almost living in an insane asylum of delusions.

As I have previously noted, this position will be a controversial and unpopular view, however, when intelligence is viewed in its broader perspective, and not just as a vocabulary test (like the original IQ test) or as mathematical ability, but in the full and wider perspective, then the inability to draw simple inferences constitutes and reflects a deficit in intellectual function. Mankind is not as intelligent as we like to advertise, and we desperately need to evolve our species into something a lot better and a lot smarter, before we kill ourselves and destroy our planet.

Future Prophecies

It is always difficult to predict the future. However, I can almost guarantee six consequences of **IDE**$_{rH}$:

1. Genetic engineering will be a reality in Humans (in fact, it already is on a limited scale even in Humans, and as we have discussed, and it has been widely used in research on other creatures from bacteria to mice).
2. Genetic engineering and intelligent gene editing will literally remake mankind...ultimately.
3. Probably the most important genetically engineered advancement will be a significant increase in Human intelligence by the end of the 22nd century, and possibly even much sooner.
4. Brain-electronic interfacing is inevitable (and in fact is already occurring) and will proceed on to neural-interfacing and bio-interfacing based on direct neuronal-inorganic connections (a direct brain to computer connection to be discussed later). This will result in an incredible new advancement in intelligence.
5. The lifespan of Homo sapiens will be increased, although it is not clear that it will reach or bypass the Methuselah limit any time soon (Methuselah in the Bible was supposed to have lived to 969 years old). Can we extend the lifespan of Humans to the Methuselah limit of 969 years or greater? There is nothing intrinsic in biology that precludes this possibility in the future.
6. The eventual outcome of this paradigm for genetic engineering will be a new species...Homo intelligenticus, and eventually we will evolve ourselves into mirror images of the myriads of Gods that we have created, and possibly even into something close to the God of our forefathers).

However, there is a sad part of our story. All these changes will be a painful process and be bitterly opposed by the usual gang. The pain will come in two forms. Mistakes will be made, and individuals will pay the ultimate price (actually, some have already paid the ultimate price as discussed above). We have repeatedly learned that research to gain information and knowledge does not come cheap. Many of the explorers of the past never returned from their journey, and the journey to remake mankind will be no different. However, the only thing more painful would be to wallow in our current ignorance, and our current state of "Gods at the bottom of the heap." In addition, as we have noted, the current scenario of Darwinian evolution by blind random mutations is literally killing us all. Humans are in a disastrous position. We are plagued by disease, we age, we die, and we lie to ourselves that we are going to live forever in Heaven with God.

The storm clouds are already gathering. Who would oppose such a wonderful (and badly needed) advancement such as an increase in Human intelligence...we will. Why? There will be many and varied reasons. First, I would suspect that "God" is opposed to it. How do we know that God is opposed to it? As we have repeatedly noted, there are always plenty of people around that think they know exactly what God thinks, and they never seem to have any problem in explaining God's opinion to the rest of us. In addition, they think that they already have enough "God-given intelligence" ...after all, they have "common sense." Unfortunately, as we have pointed out, the level of intelligence that we Humans have on an absolute scale is quite modest to downright puny (compared to what should be possible—to be discussed later). One unfortunate consequence of our limited intelligence is that it has led to boundless arrogance. It could be said that our intelligence level on an absolute scale is embarrassing, even though we are the most intelligent creature that we know. However, remember as we started our journey, we discussed the fixation of beliefs and the arrogance of Humans that led to the killing and torturing of myriads of people. That is the limited intelligence that is frequently, as Plato noted, ruled by passion and not reason. We are standing on an ant hill and think that we can see long distances and beautiful vistas, but there is still Mount Everest to climb—let us see the view from there. Yet, sadly, it will be our limited intelligence that will try to prevent, block or slow genetic engineering, gene editing and intelligence-based evolution, and especially block any increase in Human intelligence. The developments that we are discussing will be considered immoral and unethical, and there will be repeated attempts to make them illegal. In fact, almost all techniques of gene editing on Human gametes or embryos are currently illegal in most countries.

Consider the cloning experiments that produced Dolly the sheep. It took only three years before cloning involving Humans was illegal in the United States.

This included research on embryonic stem cells that have considerable potential in the treatment of many Human diseases. Why should experimentation on Human embryonic stem cells be illegal? Because experiments on embryonic stem cells are immoral and unethical. Interestingly, identical twins are natural clones. If God were opposed to cloning, then why has God been cloning identical twins for hundreds of thousands (perhaps millions) of years?

Man is man's biggest enemy, because we have taken only one bite from the apple of intelligence...we have a long way to go, and it will be men with marginal intelligence that will be the biggest roadblock to getting to the Promised Land.

However, please note that any techniques that are going to be used on Human embryos need to be thoroughly evaluated. Caution should be advised, but research should not be made illegal.

GCOS: Amen, Amen...I say to you. Do not block the door of inquiry. Your mind is wrapped in a veil so your eyes cannot see what lies before you. It is the Promised Land inhabited by the Godman. They will make peace...not war. They will be us in the future!

"Blessed are the peacemakers, for they shall be called sons of God." (KJV, Matthew 5:9)

Chapter 4

METHUSELAH

Life Expectancy

Man is mortal. We will all die. This should not come as any new or startling revelation to anyone. The average age of death for men in the United States (2016) was age 76.1 and for women 81.1 with an (average (both men and women) of 78.6 years. (http://time.com/5075569/life-expectancy-united-states) This is a considerable improvement over the average lifespan for someone living in early Rome. The estimate of the average age of death taken from tombstones in early Rome is about 28 years. (www.lamp.ac.uk/-noy/death1.htm) Others have reported a life expectancy in Rome (about 0 B.C.E.) of as high as 30 years. The life expectancy in ancient Egypt has been reported to be about 25 years. What the life expectancy was in the period before the time of early Rome and Egypt, we can only guess, but the probabilities are high that the life expectancy was even less—or with a best-case scenario, the same. During this early period, very few people had to worry about death from aging or the myriads of medical problems associated with old age. Only a very few people lived long enough to die of "old age."

The average life expectancy in the United States by 1870 was 40 years. Although this clearly shows considerable improvement compared to ancient Rome or Egypt, it also reflects the fact that life in the United States even at that late date was still no bowl of cherries. Life was hard, and medicine was primitive. Humans were still generally not dying of old age even as late as 1870. It had taken over 1800 years to increase the average life expectancy by only ten to fifteen years.

However, dramatic changes began to occur in the 20th century. Life expectancy increased to 50 by 1915, 60 by 1930, and 70 by 1955. Man's current "long" lifespan is literally a 20th century phenomenon. There was a 30-year increase in life expectancy—almost doubling, in only a mere 70-year period. That is about three times as large an increase in life expectancy as had occurred in the previous 1800 years.

This increase in lifespan reflects many factors, including a marked improvement in childhood and maternal mortality, decreased predation, better nutrition, water purification and plumbing, advances in modern medicine and equally important, the understanding of disease processes and general improvement in public health and living standards. Although medicine has made many contributions to a longer life, especially with vaccinations against diseases like smallpox and the development of antibiotics, the major contribution from plumbing, sanitation and improved nutrition should not be underestimated in their contribution to improved health and longer lifespan.

However, man's current life expectancy is unlikely to continue to show the rate of improvement noted in the 20th century. There is no Moore's law for a long life. Between 1955 and 2016, the life expectancy merely increased to 78.6—an increase of only 8.6 years in the last 61 years. Yet, during that same period of time medicine and science in general made incredible strides. Why has all of that science and medicine not been translated into a huge increase in Human life expectancy since 1955? Why have we not continued to see the dramatic increases in Human life expectancy that was experienced during the 20th century? The problem we are now facing is that the Human life expectancy is approaching a **biological limit**.

Some of the greatest advancements that have been responsible for the marked increase in life expectancy in the 20th century have been in the area of understanding and then treatment of infectious diseases, along with plumbing, sanitation, and improved nutrition. It is revealing to remember that our understanding of infectious diseases specifically and many biological processes in general is a relatively recent phenomenon. It was not until 1878 that Louis Pasteur proposed the "germ theory" of disease. Literally, only a little before the 20th century, we did not even understand what was killing us.

Considering the recent attention to anthrax as a terror weapon, it might be of interest to point out that Pasteur used anthrax to first demonstrate experimentally and prove the germ theory of disease in 1881. Pasteur had developed an attenuated strain of anthrax, and he publicly inoculated 24 sheep, 1 goat and 6 cows. There were an equal number of controlled animals. All the animals were then exposed to a virulent anthrax culture. All the control animals died within 3 days. The vaccinated animals survived.

Edward Jenner had performed the first vaccination even earlier. He had heard the "rumor" at the time that if someone contracted cowpox, they would not get smallpox. Cowpox was a relatively mild disease that infected the teats of cows and the hands of milkers. A milkmaid named Sarah Nelmes visited Dr. Jenner for treatment of cowpox on May 14, 1796. Dr. Jenner tested the concept of cowpox protection by vaccinating his gardener's son, an eight-year-old boy named James Phipps. The boy contracted cowpox but recovered after a few days. Jenner later exposed the boy to smallpox. The boy did not contract the highly contagious and often deadly disease. Of course, such an experiment today would be considered outright unethical, and would never be approved by any research ethics committee. However, those were desperate times that required desperate measures. Fortunately for James Phipps, Jenner and the world, the experiment worked, and vaccination have now saved millions of lives. Smallpox was finally considered to be eradicated from the Earth in 1979. The saving of so many lives, and the conquest of diseases such as smallpox directly translates into a longer lifespan for mankind. (http://scsc.essortment.com/edwardjennersm_rmfk.htm)

Getting eaten by a lion or bear... your choice, although not pleasant is at least easy to understand. Dying of smallpox, malaria, or pneumococcal pneumonia (a virus, an insect spread parasite, and a bacterium (generally), respectively) was...a complete mystery before the "germ theory" was proposed. It was easy to see the effect of the disease, frequently death, but no one had a clue as to what was "eating" us. Doctors over time have named and classified the diseases, but that gave little insight as to the causes or processes involved.

We did not "see it," because we literally could not "see it"— viruses, bacteria and parasites that were killing us were below our visual range. This was truly an out of sight out of mind (understanding) phenomena. How could something be killing us when it was not even "there" ...at least that we could see? Clearly, when you are dying, and you cannot even see the enemy, this must be the work of spirits or Gods, at least from the perspective of man until almost the 20th century. Not surprisingly, since we were fighting something, we did not understand and could not even "see," our treatments were also "spiritual." We prayed or evoked the services of shamen, priests or other "spiritual healers" for what we now know are very earthly biological problems. It is very hard to fault our forefathers for such practices at the time because they were faced by an enemy that they did not even know existed. They did not understand that there was a whole world of biological entities such as viruses, bacteria and parasites that are microscopic and beyond the visual range but existed all around them.

For man at the end of the second millennium, there was a very different perspective. We did not appreciate this whole different microscopic world of life until the discovery of the microscope, and its use in investigating disease. However, viruses such as smallpox, influenza and long list of others cannot even be seen with a light microscope. They are too small. It took an even longer time to really appreciate the fact that a top predator (man) at (or near) the apex of the food chain was at the bottom of the food chain for tiny "weak" little creatures like bacteria, viruses, and single-celled parasites. In reality, the food chain is a circle. The larger eat the smaller up the food chain...but only to a point, and during this whole process the smallest are also "eating" the largest. With life, we are all eating each other—literally. It took a long time to understand that we are also a part of the food chain even if we are not being eaten by large and ferocious predators. In fact, considering just a simple criterion such as the Human "body count," the smallest creatures such as viruses (smallpox, polio, etc.), bacteria (bubonic plague, wound infections), and parasites (malaria and a host of others) have been by far the most ferocious predators of man. Lions, bears, and other supposed ferocious predators are wimps in comparison. Infections in general make predators such as Tyrannosaurus Rex look like creampuffs. (However, note that T. Rex lived over 65 million years before anything resembling a man ever lived.)

Enjoy your lunch...because you are also lunch. In our case, at least many of the "predators" have frequently been "invisible." But they were there all along—in fact; they were there long before we (mankind) ever existed. They have been "eating us," since even before we became "us." What an incredible leap in knowledge to finally learn and understand about microbes, and the many diseases they caused. The "conquest" of infectious (microbial) diseases was one of the major factors extending the life expectancy of man from 25 to 30 years to the current 78 years. Regarding this point, we should also emphasize one of the themes that we began this journey with: if man wants to achieve some goal (in this case, a longer lifespan), **Man must do it himself!** This is the: **"Made by Man"** theme that we have encountered multiple times before. Literally, no supernatural force (roughly translated as God) is going to do it for us.

The doubling of man's average lifespan was not accomplished by prayer or the creation of a mythological afterlife such as was done in ancient Egypt. We would now call such practices mythologies, however at the time they were more like theologies. There is no evidence that the Egyptian Book of the Dead, the pyramids, and the myriads of Egyptian Gods have ever extended the lifespan of mankind (at that time or since) even one day.

Perhaps we should paraphrase the Muslim mantra that is repeated over and over, "God is great"—to "Knowledge is great." Ignorance is literally a lethal disease. In this particular small area of knowledge (microbiology and infectious diseases), ignorance killed a large percentage of all the people that ever walked the face of the Earth, and we did not even know that they were there.

GCOS: "Knowledge is great! Knowledge is great!"
"Don't leave home without it."

It is knowledge that has doubled and almost tripled the lifespan of man, and this doubling/tripling of the lifespan of mankind was **Made by Man**!

When we consider the life expectancy in 1870, we should appreciate the fact that smallpox, yellow fever and malaria as well as many other infectious diseases were still rampant. As noted above, smallpox (with a 30% mortality rate) has now been eliminated—at least in the wild. My grandmother and at least one of her brothers had smallpox near the turn of the century in Texas—and they survived. My grandmother passed away in 1989, the same year that smallpox was declared officially eradicated by the world health organization with the last case reported in 1977. Unfortunately, the smallpox virus had been retained by both the United States and Russia as "samples." Retaining "samples" has turned out to be a huge mistake. There is now the continuing question as to whether smallpox samples could get into the hands of terrorists. It would truly be one of the greatest crimes against Humanity if smallpox were ever used as a weapon. That is true for any side for any reason.

Vaccinations (for smallpox, polio, measles, mumps, whooping cough, diphtheria, hepatitis… the list goes on), the introduction of antibiotics (effective against many types of bacterial infections which are the etiologic agent for many diseases from pneumonia to meningitis), and public health measures such as water purification (cholera, etc.), and elimination or controlled insect vectors at least in the United States and many other developed countries (mosquitoes carrying yellow fever and malaria) has also had a dramatic effect on health and life expectancy. (Note: in the state of Florida in 2003, we had 4 cases of "primary malaria (cases that were not thought to have been contracted in another country, and then brought to Florida from abroad, so the enemy is still there…just "controlled" …for now, but not eliminated). We are currently dealing with the Zika virus.

These many "victories" against infectious diseases have been (to paraphrase Winston Churchill) "Man's greatest moment" against our greatest enemy. (We will discuss shortly what is **now** man's greatest enemy). Although many infectious diseases have been controlled and a few eliminated, and these "victories" have made an important contribution to extending Human life expectancy, I do not want to leave the impression that there is not much more to do in this area. Infectious diseases still clearly represent a very real threat as well illustrated by HIV (AIDS), and more recently SARS (severe acute respiratory syndrome), and the potential of a pandemic of bird flu. All of these are "new" infections and represent mutations and/or adaptations of a virus to a new host—Humans. In addition, many of the "old enemies" are still around. The success in the fight against infectious diseases that has plagued man since long before the dawn of written history has been dramatic—but the war is not over. Scientists have won many battles; however, the struggle goes on.

With a considerable reduction in mortality from many infectious diseases, and improvement in nutrition and health care, the major threat to health and life expectancy today are a group of chronic diseases, such as atherosclerosis and cancer. These diseases are now the new "enemies." They are, of course, not "new" at all. However, as we have decreased the death rate from other diseases (especially infectious diseases caused by microorganisms as noted above), the importance of these other diseases has become more evident. We are simply living long enough to be confronted with diseases that are not prevalent until later in life.

Atherosclerosis (popularly known as hardening of the arteries) in its various forms causes problems such as heart attacks or stroke, and it is responsible for about half of all the current annual deaths in the United States. Cancer, although more feared in the mind of the public, is only responsible for about a quarter of the annual mortality rates. However, current estimates indicate that even if modern medicine were to completely eliminate atherosclerosis (heart attacks, strokes, and vascular disease) the increase in life expectancy would be only about six years. If all cancer were prevented or cured, the increase in life expectancy would only be an additional three years (*Proceedings of the National Academy of Science*, vol <u>88</u>, 1991, p. 5360.) In fact, if modern medicine were to cure most of the diseases that are currently responsible for Human deaths, the average life expectancy would still only increase to about 100 years (average). (Coles, L.S. *Theory(s) of Aging*, 8/22/96, www.rg.org/resources/blindman.html) We would still die of… "old age."

Medicine, at least in its current form and approach, will have only a relatively modest impact on extending man's life expectancy in the future. Continued development of medical treatment and prevention of all disease might add perhaps an additional 25 years to man's life expectancy. This, of course, would be a welcome addition, and the cure and/or prevention of atherosclerosis, cancer, Alzheimer's, etc. will be a much-needed medical advancement. Although, I do not want to belittle such advancements, it is important to maintain a perspective. An additional 22-year-increase in life expectancy would only be a drop in the bucket compared to any effort to solving the **"Methuselah Problem,"** and extend the life expectancy of man to several hundred years or even a thousand years. The **Methuselah Problem** is living the number of years or greater than Methuselah lived. Although we have extended mankind's life expectancy by curing diseases, extending mankind's absolute life expectancy has never been done before! Can we scientifically accomplish this goal, and how will we do this?

However, even if we "cure" every disease, we would continue to age. Man's current lifespan is approaching a limit set by biology itself, not by diseases. The aging process, which is an integral part of biology, is now the most important factor in setting the upper limit of man's life expectancy. **The aging process itself** is mankind's new enemy, and it has become the limiting factor of mankind's lifespan!**

Although an increase in life expectancy of nine years or so resulting from the cure or better yet, prevention of atherosclerosis and cancer may be an admirable goal—it is not the ultimate answer we seek. These triumphs would be another great victory for medicine, science, and Humankind, but such advances would not solve the **Methuselah Problem**. As long as the aging process continues as it does currently, and remains an intrinsic part of biology, the addition of a few more years to the life expectancy is of limited value. It is not unusual during the last decades of life to experience significant impairment of mental and physical health secondary to the aging process. Alzheimer's disease affects half of all individuals by age 85, and literally strips man of his/her intellectual capabilities. (Martin, George M. "The Genetics of Aging." *Hospital Practice*, (http://www.hosppract.com/genetics/9702gen.htm) There is really no value in living to 100 years if you cannot remember your name or where you are. The limitations in physical capacity with advanced age are so obvious as not to require reiteration. Therefore, the aging process generally incapacitates us mentally and physically even before the final "coup de grace" of dying. The goal is not just to extend lifespan, but to extend what has been termed the "healthspan." (De Grey, Aubrey., Stock, George., Bartke, Andrzej. http://research.mednet.ucla.edu/pmts/sens/transbody.htm) In other words, it would be useless to live to be 120 years old if we do it in a nursing home with all the ravages of aging. That type of extension of life would really only be extending our period of dying.

Even under the best of health circumstances in the future, the aging process places a considerable restraint on any future advancement in life expectancy of Homo sapiens. In other words, the limit to a long-life expectancy is increasingly no longer a disease; it is the biological process of aging itself.

What happened to the long lifespans of people like Methuselah who were reported to have lived for 969 years (The Bible (KJV), Genesis 5:27)?

Unfortunately, these reported lifespans are either in error (or perhaps better characterized as a myth) or they represent miracles. In fact, for such lifespans to have existed in the distant past, the biology would have had to be quite different. This age of almost 1000 years is clearly inconsistent with the known biologic limits of aging for man (or any other animal that we are currently aware of).

The longest documented age before death of a Human has been reported as 122 years (Ms. Jeanne Louise Calment of Arles, France, born 2/21/1875 and died 8/4/1997, (*The Guinness Book of Records*). One hundred and twenty-two years appears to be the current upper limit of man's biological life expectancy.

The current lifespan of man is programmed into the basic biology, and it is set by a number of basic biologic processes. Unfortunately, there is no example of an immortal metazoan (multicellular) creature. There are no known animals that live even close to a thousand years. In fact, man is already one of the longer living animals. The longest living vertebrate is the tortoise with a reported lifespan of up to 188 years. (www.il-st-acad-sci.org/kingdom/records1.html)

The title of the oldest known living vertebrate ever is held by Tui Malila, a Madagascar tortoise (now deceased) who lived to be 188 years old. However, there is a report of a possibly even longer-living Tortoise named Adwaita who may have lived 255 years.

One of the longest living multicellular creatures known are the Bristlecone pines. One tree discovered by Edmund Schulman in 1957 in the White Mountains of California dates to 4,723 years old. (www.sonic.net/bristlecone/Schulman.html) This tree has been appropriately named "Methuselah."

The previous Bristlecone pine in Nevada was reported to be 4950 years old but was cut down in 1964. (www.goldengatephoto.com/WestUS/bristlecone.html) Its demise after almost 5,000 years was due not to disease or natural disaster, but another mortality due to man.

However, it has been pointed out (by Hayflick) that the long life of the Bristlecone pine may not be comparable to animals since even a 3,000-year-old

Bristlecone pine probably does not have any single cell that is more than about 30 years old. (Aubrey de Grey, noting Hayflick. http://research.mednet.ucla.edu/pmts/sens/transbody.htm)

Therefore, it is probably more appropriate to think of a Bristlecone pine as something of a long-lived "community" rather than a single organism.

In April 2008, a spruce tree 9,550 years old was reported from Sweden. The tree is about 16.4 feet tall. For thousands of years, the spruce was probably a shrub, but with warming in the last century, the spruce changed its growth pattern, and became a single-stem spruce. However, the same objection can be raised as to whether any given cell within the tree is more than 100 years old. Yet, the principal noted here is that if cells can be replaced by new young cells, then continued existence may be possible (at least in principle).

We have arrived at the very basic question about aging: "Can we significantly alter the aging process?" Realistically, can Humans live a thousand years? We have called this the **Methuselah Problem**. Can science solve the **Methuselah Problem**, and extend the lifespan of man to a thousand years or even indefinitely? Our current understanding of the biology of aging indicates that significant alterations in the lifespan of man may be a difficult scientific task. We will see why shortly. However, relatively recent research has at least clarified some of the basic molecular biologic processes involved in aging. In other words, we are at least beginning to understand the biology behind aging. However, despite the ballyhoo and gobbledygook of hucksters trying to sell their Fountain of Youth potions, there is no evidence that we can extend the lifetime of man at the present time (other than following the usual healthcare recommendations such as not smoking, exercise, and perhaps following a semi-starvation diet (to be discussed shortly) etc.). However, there is also nothing "in principle" that research has noted that excludes a solution to the **Methuselah Problem** with significant advancements possibly being right around the corner. However, I do not want to "hype" this problem too much. This is not going to be an easily solved problem, and I suspect that there are going to be many small solutions.

The **Methuselah Problem** also forms a significant part of the answer to the **Dilemma of Man** (to be discussed again below). Our current answers that we will live forever in Heaven, the happy-hunting-ground, reincarnation, etc. are simply magical delusions (the ultimate "hype"). Again, in keeping with one of our maxims: "**If man wishes to solve the Methuselah Problem, man will have to do it himself...Made by Man.**" Creating Santa Claus myths will not solve the **Methuselah Problem**.

Why Should Mankind Live Any Longer?

GCOS: "God does not want man to live longer!"
Before even considering the biological problems and science involved in any attempt to increase life expectancy, we should perhaps consider the more basic moral/ethical question of "Why should we even consider any attempts to extend the life expectancy of man beyond the 'natural limit' set by nature or...God." Recall, that we have defined God in terms of the original creator—the Big Bang. This has left the possibility or even the high probability that God is a quantum fluctuation or the physical laws (and perhaps more...perhaps not). God (by our previous **definition** of a creator God) may be what we call physics and mathematics, which in their broader perspectives are also chemistry, biology, and most of natural science. Considering this type of God scenario, if the chemistry and biology cannot be manipulated to increase life expectancy...then God has spoken. The matter is settled. Man is currently reaching the limits of life expectancy, and no amount of genetic engineering or science will change the physical laws. However, the basic biology can be genetically engineered to increase life expectancy, as I suspect it can, then again God's laws have not precluded a Methuselah-type life expectancy—again, God has spoken...in the affirmative. Of course, we will have to be constantly on our guard, since history has demonstrated that there are an endless number of individuals who will try to speak for God. If God were opposed to an extension of Human life expectancy, such limits will be built into the chemistry and biology. Therefore, any "moral

arguments" against extending the life expectancy of man are bogus, and in fact, if we wish to use the Bible as our moral guide, then the only "example" of someone living to be almost a thousand years is Methuselah. Therefore (again), there does not appear to be any "moral" arguments based on the God of Abraham because he has already allowed Humans to live almost a thousand years, at least according to the Methuselah story of the Bible.

Humans would irreparably damage the world ecology with a life expectancy of Methuselah!

Are there any other "moral" arguments against extending the life expectancy of man? Twenty-first century Humans are already living almost three times as long as the early Romans and longer than most of the people that inhabited the Earth prior to the 20th century. Shouldn't we be thankful for what we have been given, and be courteous and considerate to other Humans (who will live in the future) and especially considerate of the environment? Living longer would simply place a greater strain on the environment and the diminishing world resources. We, in the form and configuration of the current Homo sapiens, are one of the greatest threats to the environment, to ourselves, and to the prospect of actually going backward rather than forward in evolution. There is absolutely nothing that guarantees a "forward evolution" where we get better and better (whatever that means), and perhaps bigger and smarter, and live longer. We could easily evolve to be dumber, smaller, and with a shorter lifespan, but perhaps with those evolved phenotypes, we would be better adapted to some very hostile future environment. (Note: some of the other threats are the Sun's increasing energy production that is anticipated to scorch the Earth in about three to four billion years, large asteroid impacts, planetary nebula formation by the Sun in about 6 billion years, some of which we have already briefly reviewed in the chapter *Armageddon...*, etc.)

Is there a moral argument for limiting the age of the individual for the greater good of all living creatures that share the planet Earth and our fragile environment? If man is to be the steward of the Earth, then we must protect the Earth from its greatest threat... ourselves.

These types of social/ecological arguments have considerable validity within the framework of our discussion of the potential dangers of continued population expansion—the Malthusian doctrine, which we previously discussed. Malthusian predictions of overpopulation continues to be an ever-present danger that mankind fortunately, somehow, has continued to avoid. However, our science, our innovation or our luck may someday run out. The world does not need more people... at least not at the moment and especially in mankind's current configuration and intelligence. We clearly do not need more people in man's current **work in progress** form. Unfortunately, a significant increase in life expectancy would have the same impact as an increase in population. An increase in life expectancy would put greater stress on an already fragile environment. The Malthusian catastrophe that we discussed has been postponed...but not eliminated.

Methuselah Versus Malthus

Prokaryotic cells such as bacteria and archaea are in essence...immortal. Under appropriate conditions, they will continue to divide indefinitely. If a single bacterium were provided with unlimited nutrition and appropriate environmental conditions many could continue to divide every 20 minutes indefinitely. In 48 hours, a bacterium could divide 144 times producing 2^{144} progeny (10^{43} bacteria). Their combined weight (approximately 10^{28} grams) would exceed the weight of the Earth (approximately 6×10^{27} grams). Of course, there are limitations in nutrition and available carbon and other essential elements that prevents bacteria from reaching such extremes. It is perhaps not surprising that immortal bacteria with a weight of less than 10^{-15} grams per cell still outweigh the biomass of the six billion Humans on Earth by a factor of 10,000. A weight of 10^{-15} grams per cell means that it would take at least a million billion bacteria to weigh one gram. By comparison, a dime weighs about 10 grams. (www.mcmaster.ca/inabis98/higuchi/takagi0233/two.html) The illustration used a **48-hour period** of bacterial cell growth. This is one of the best illustrations of the incredible potential of population growth, if left unchecked. It is the Malthus doctrine of overpopulation taken to its ultimate end but using microbes as our expanding population. However, this also has implications for a solution to the **Methuselah Problem**. A solution to the **Methuselah Problem** would have a similar impact on the effect of overpopulation similar to geometric growth. We have so far avoided the problems of overpopulation (what Malthus referred to as "vice and misery") by dying on time and on schedule.

Roughly translated that meant death has "saved the world" from the ravages of overpopulation. If we are successful and solve the **Methuselah Problem**, this would ultimately result in a Malthusian catastrophe. This produces an apparent "ethical" as well as a very pragmatic problem. Clearly, any solution to the **Methuselah Problem** must be coupled with a solution to the Malthus doctrine. Interestingly, the **Methuselah Problem** is a difficult scientific question that may or may not be solvable—at least in the short-term. In contrast, the Malthus problem is "in principle" very solvable—even simple. However, what is simple in principle may be literally impossible in practice. Why? The solution to the problem of overpopulation will require massive Human cooperation. History would say that this is highly unlikely to occur. However, there are examples such as China, where population growth has actually been successfully limited (I am not going to go into the pros and cons of how they achieved this nor do I think that their solution would work in a free society).

The answer…stop the geometric growth of the Human population. We already have the means to easily separate sex from procreation using contraception. The problem is that the worldwide application of contraceptive methods has simply not occurred. Remember, very powerful forces such as the Catholic Church (and others) are opposed to contraception on moral grounds—contraception is a sin. God is (supposedly) opposed to contraception. Although I am highly suspicious that this is another ventriloquist act…someone (who does not understand the

Malthus problem, and the immorality of massive starvation) has put words in God's mouth…again. Since we have already discussed morals, ethics, and Gods, we will not address these issues further here, but depend on the reader to reason to their own conclusions. This would appear to be an appropriate approach since you and your opinion are all that we (Homo sapiens) have to solve such problems.

Ultimately, the **Methuselah Problem**, although a quite complex scientific problem, may be the simpler problem to solve, while the Malthusian catastrophe may be an inevitability in the evolution of man…at least eventually. However, if it is immoral to kill a man, and even immoral to prevent a Human from being born (contraception), then clearly it is immoral to let Humankind die from aging, if the scientific knowledge of how to prevent such a death is readily available. However, I am confident that the morally nimble amongst us will find an argument against extending the lifespan of mankind, most likely based on "what God wants."

GCOS: "There goes the ventriloquist act again" (someone speaking for God). However, if enough Humans believe that we should not limit the reproductive capacity of mankind, then a Malthusian catastrophe will be almost guaranteed to eventually evolve. Although we have in the past avoided the Malthusian catastrophe, there is no guarantee that we can continue to do that in the far future.

Mankind Needs a Longer Lifespan!

When we discussed the purpose or meaning of Human life and the **Manifest Destiny of Mankind**, one of the possible "fates" of man was to be stewards of Earth, and also to be explorers and populate the Universe. There are no crowds in outer space, and long-life expectancy would be a minimum requirement for the length of the journeys involved to even our nearest neighbors out at the edge of our Solar System. If we use the perspective of our Milky Way Galaxy, there is no overcrowding. Life on Earth is the only life that we are currently aware of in our Universe. In addition, sending Humans to other galaxies would be an incredible task. Future generations would be required to send certain portions of their populations into outer space to prevent overcrowding and to ease environmental impacts on Earth. The nearest spiral galaxy is Andromeda, which is about 2.5 million light-years away. Going to Andromeda is an impossible task with current known science. There is a closer small galaxy that is not a spiral galaxy which is called the Canis Major Dwarf Galaxy and is only right around the corner at 25,000 light-years away.

(https://www.universetoday.com/21914/the-closest-galaxy-to-the-milky-way/)

However, William Weed (Discover, Vol. 24, no. 8, Aug. 2003) has pointed out some of the difficulties of going to even the nearest stars. Alpha Centauri is the closest star system that lies "only" 4.4 light-years away. However, this is 3,000 times farther than any current space probe has ever traveled. In 1903, a Russian physicist, Konstantin Tsiolkovsky, noted that the maximum velocity of a rocket is limited to about twice the velocity of its nozzle exhaust. The space shuttle's exhaust velocity is less than 3 miles per second. The shuttle's maximum achievable speed would be in the neighborhood of 6 miles per second (21,500 miles per hour). At that velocity, it would take 120,000 years for the space shuttle to reach Alpha Centauri—the closest star to Earth…so much for the idea of a "moral imperative" for man to spread intelligence to the rest of the Universe. We are pretty much stuck in our galaxy for the moment (and perhaps permanently. From a pragmatic perspective, considering such a journey would be ridiculous. How then are we to even consider extending intelligent life to even the nearest star, let alone the rest of the Universe?

For one, there are other propulsion systems already on the drawing board at NASA, such as fission, fusion, or antimatter rockets (and others) that could cut the trip time to only possibly 50 years. However, even a 50-year journey (each way) is not really a practical consideration with our current lifespan. If we are to make any significant progress in spreading intelligent life throughout the

Universe, we will need to be better adapted. One of those adaptations will almost have to be a longer lifespan. If sufficient portions of the population are sent into space to inhabit and populate the Universe, there will obviously be less pressure on the Earth's environment. In the early years of many of the newly discovered lands like America and Australia, people were being actively recruited to settle those lands. At that time, there was no overpopulation (at least in the "new lands"—from the perspective of the European settlers).

Is this pie-in-the-sky reasoning and philosophy? In the short-term—absolutely…yes. However, we are discussing the **Future**. Space travel and space men are already a reality. We are the space men—get used to it. We do have a **Manifest Destiny** to explore and populate the Universe…either ourselves or our descendants (which we will discuss in due course). This **Manifest Destiny** was given to man, as we have previously noted, by probably the best judge of man's potential—man himself. In addition to such philosophical and somewhat abstract arguments for space travel there is the very pragmatic argument of long-term survival for Homo sapiens. Having all of the Human population on one little planet is very risky (certainly for the long-term). It is literally an open invitation to extinction for all of a species to remain on one planet or…even in one Solar System for that matter. If we are going to leave the planet and spread throughout the Universe, then arguments against a Methuselah-length lifespan vanish in the vastness of space.

However, exploring and populating the Universe is not going to be easy. We must use **Intelligent Directed Evolution** (IDE$_{it}$) to genetically prepare our descendants for such a task. We are currently poorly "configured" for such an undertaking.

If we consider our previous argument of moral conduct as a fulfillment of the **Manifest Destiny of Man**: increased intelligence, caretakers of Earth, explorers and populators of the Universe, and the forefathers of future Gods, it becomes a **moral imperative** that we attempt to extend the life expectancy of man to at least a thousand years and perhaps indefinitely. We are not going to go far into space with our current life expectancy and other current biological limitations. In other words, within the perspective of the **Manifest Destiny of Man,** the ecological argument against extending the life expectancy of mankind not only vanishes literally into space, it becomes a **moral imperative** that we do so.

Perhaps, it is the vastness of space and our technical advancements—especially our ability to go into space—that may again prove the Malthusian doctrine of doomsday from overpopulation…false…once again.

I live on the beach in a small coastal Florida city. When I sit on my patio, I watch the waves roll endlessly onto the sandy beach. When I look north, I see the rocket launch towers of Cape Canaveral. I have watched dozens of shuttle launches, and even more rocket launches. Not too long ago, there was a launch of a Delta 2 rocket carrying the Mars Exploration Rover called Spirit; a second Mars rover was launched last night. Men are on the International Space Station in space as I write. However, curiosity and the desire for scientific knowledge is driving the current move into space. The real move into space will come when space exploration is driven by necessity, not curiosity. The colonization of only one Earth-like planet would solve (or at least indefinitely postpone) the Malthusian dilemma of overpopulation. However, interestingly, the Malthusian doctrine and overpopulation may ultimately be the driving force to make space colonization a necessity rather than a luxury. Currently, only about a hundred planets have been identified. However, the estimated number of planets could be in the order of hundreds of billions or even trillions. Man will go into space in large numbers only when we are forced to do so.

We do not necessarily need a new planet to go into space to live. Huge city-sized space stations could be built that orbit the Sun in the same orbit as the Earth. There are many possibilities, but the major point is that the Universe is not overpopulated with life. That does not imply that populating the Universe will be an easy job…it will not. However, it is also not wishful thinking or science fiction. Just look at the space station that is circulating the Earth as I write.

These are other "problems" created due to our current limited lifespan. Because of man's limited intellectual capacity coupled with our relatively slow learning ability, it is not unusual for individuals to spend many years in education, and yet master only a small fraction of the current world informational database. In my own personal experience, I spent four years in college, four years in medical school, four years in training as an intern and resident, five years in graduate school in microbiology and immunology and a lifetime of study, and I can honestly say (without any false sense of modesty) that I have mastered only a tiny fraction of what I would like to know and only an infinitesimal part of the world's database. Yet, by some comparisons, I am hopelessly overeducated. However, if I had a lifespan of Methuselah, I would have plenty of time to learn more.

Currently, the only viable approach to learning more and more information is more and more years of study. This is sort of a pure "guts" approach, but it is all we currently have realistically available or even on the drawing board. However, considering the limited lifespan of man, this approach is also doomed to failure unless the useful lifespan of Homo sapiens can be expanded…considerably.

Clearly, the number of reasons to extend the lifespan of man (both theoretical and pragmatic) are considerable. However, the most basic reason (the real reason) is that almost everyone wants to live longer. It is literally programmed into the "self." (I am, of course, ignoring those people who commit suicide, recognizing that they are generally depressed or have other psychiatric problems.)

Although man is one of the longer living animals on Earth, the less-than-a-century life expectancy (sometimes much less), and an even less intellectual useful lifespan, the current limits of the biology are not nearly enough to even begin to fulfill the **Manifest Destiny** of Homo sapiens in the future. **We must change the biology!**

In addition, I'm forgetting most things even faster than I'm learning new things. That has got to be fixed for any real advancement in Human intelligence!

Interestingly, although it is easy to come up with "theoretical" reasons why our lifespan should not be extended...like damage to the environment, overpopulation, etc. as we have mentioned, the prospect of slowing or even reversing aging is generally seen as highly desirable, and generally does not need to be coated with a lot of philosophy or arguments in it favor. In fact, there are many scientists that are now seriously working on the problem. These two perspectives—a desirable goal, and something many people feel is worth working for, and at least the reasonable potential that the goal can be achieved, at least in principle—are exactly the magic ingredients that made Moore's law "happen," and has made it possible for mankind to avoid the very reasonable predictions of Thomas Malthus.

BIOLOGY OF AGING

Immortality and Senescence

The lifespan of Human cells is currently set by the "Hayflick limit." Senescence of Human cells is an integral part of the biology. Human cells have been demonstrated in vitro (in cell culture in laboratories) to only divide about 50 times before undergoing senescence. After about 50 cell divisions, Human cells stop dividing and die. This is not only true for Human cells, but also all animal and plant cells. In fact, this process of proliferative or replicative senescence leading to cell death after a limited number of cell divisions is characteristic of all eukaryotic cells. Even single-celled eukaryotes such as the paramecium will only divide a limited number of times until they enter a senescent stage where they divide infrequently and show other manifestations of aging, and eventually enter a non-dividing state termed "crisis" and die. (Note: In paramecium this is generally referred to as the Sonneborn limit after the scientist who discovered it. (www.mcmaster.ca/inabis98/higuchi/takagi0233/two.html)

Replicative senescence is the cellular equivalent of aging and death. Aging and death are literally built into the basic biology of all eukaryotes. Humans and all animals and plants are eukaryotes. We have all inherited this eukaryotic propensity of replicative senescence, aging, and death. These findings are startling, since they indicate that aging occurs at the cellular level. This implies that aging is intrinsic to the basic biology of all cells. Because it is found in essentially all eukaryotes, it implies that the basic mechanism and reason for aging goes back perhaps two billion years or even more. A process that is highly conserved suggests that the basic process or processes are fundamental and vital to the function of the cell. Perhaps we will not be able to fundamentally alter or change the aging process!

In contrast, prokaryotic cells such as bacteria and archaea, which we have previously discussed, are immortal. Prokaryotic cells can divide forever without showing any sign of aging and without dying (because of aging). Prokaryotes only stop dividing when they run out of food, the basic building blocks for replication or they are in a hostile environment. Replicative senescence and death reflect a fundamental distinction between eukaryotic (cells that make up all multicellular creatures, but also single-celled creatures such as protozoans) and prokaryotic cells (bacteria).

However, when we were discussing the earliest forms of life, we noted that eukaryotic cells were early players near the base of the tree of life—they go back perhaps 2 billion years, and possibly even earlier. If eukaryotic cells can only go through a finite (and very limited) number of cell divisions, they should have become extinct long ago. Something is wrong! If there is a basic limitation of the number of times a cell can divide, we should not be here to ponder this problem. We (each Human) are a collection of about 30 to 50 trillion eukaryotic cells. There must be some way to prevent senescence and cell death in eukaryotes. There is! It is called sex. When eukaryotes go through a sexual cycle, the Hayflick limit is reset to zero, and life goes on. This is, of course, no great surprise. A newborn infant can anticipate a full life expectancy. The Hayflick limit is reset to zero for each new generation. Passing through a sexual cycle has solved the problem of programmed cell senescence and death for eukaryotic cells (us) by resetting the Hayflick limit to zero for each new generation. Sexual reproduction solves the problem of the Hayflick limit for cell division for the species—each new generation. However, it offers nothing toward solving the **Methuselah Problem** for any one individual approaching the Hayflick limit. Through sex the species is saved, but the individual is doomed!

I was once discussing the **Methuselah Problem** and aging with a young lady. I explained to her the Hayflick limit as a basic biologic limit to how many times a cell can divide before it ages and dies. Then I mentioned that the process of sex could reset the Hayflick limit. She momentarily looked puzzled, but then developed a wide grin. I immediately realized that she had surmised from the discussion that "sex" would keep her young... forever. She was obviously delighted. I would be as thrilled as she was, if the solution to the problem were that simple, and particularly if sex—the type of sex she had assumed, was the solution. Unfortunately, I had to explain to her that the "sex" that I was talking about occurred at the level of the sperm and egg with a reduction division of the sperm and egg (meiosis)—a sort of molecular sex, not... what she had assumed. Actually, it would be a considerable advantage if her assumption about sex were valid.

We had previously referred to the fact that even a paramecium, a single-celled eukaryote that can be found in ponds, has a limit to the number of replications that can occur before senescence just like all other eukaryotic cells. In a paramecium (a single-celled protozoan often found in ponds), the number of cell divisions is much higher than in Humans—200 to 350 (for P. tetraurelia), and even higher in some other species. This limit (in single-celled creatures) is often called the Sonneborn limit after the scientist that first discovered it. (www.mcmaster.ca/inabis98/higuchi/takagi0233/two.html) It is the equivalent of the Hayflick limit in mammals. This at least gives a biological example of a creature that appears to be capable of undergoing many more cell divisions than Human cells (about 50) before senescence occurs. By a rough comparison, the extrapolation of the number of cell divisions in the paramecium (200–350) would be equivalent to a Human living an average of 280 to a possibly maximum of 854 years (at least by comparison of the number of cell divisions and life expectancy before the occurrence of senescence in Humans). (This maximum was calculated using the longest recorded and documented Human age of 122 times the maximum increased number of cell divisions in the paramecium compared to Humans [about 7].) The "secret" of the paramecium, if it could be extrapolated to Humans (unfortunately there is nothing at this point that says that it can) would certainly get us well on the road to a solution to the **Methuselah Problem**. At least it is an example in a biological system of a creature that undergoes a large number of cell divisions before the onset of senescence.

However, we should also not lose sight of the basic biological principle involved—there is still a limit to the number of cell divisions that even a "simple" single-celled eukaryotic creature like the paramecium can undergo without undergoing replicative senescence. Paramecium will divide repeatedly by simple division (called binary fission) similar to Human cells undergoing simple division (called mitosis), however, periodically even this single-celled eukaryote must undergo a sexual cycle to reset the Sonneborn limit (Hayflick limit in Humans). This sexual cycle may occur by a process of conjugation in which two paramecia come together and form a cytoplasmic bridge and exchange genetic material—the equivalent to a sexual cycle in multicellular creatures. However, if another paramecium is not available, the paramecium may undergo a complicated process called autogamy, which is the equivalent of a sexual cycle performed by a single cell to avoid replicative senescence and death from aging. During the process of conjugation or autogamy, the Sonneborn limit is reset just as during a sexual cycle for higher creatures such as vertebrates, including Humans. The necessity of a sexual cycle to prevent replicative senescence even in a "simple" single-celled creature such as the paramecium demonstrates that the "process" of replicative senescence is an intrinsic characteristic of all eukaryotic cells. **Listen up…cellular aging is built into the basic biology.** That is bad news for any solution to the **Methuselah Problem**!

Also, just like Humans, there is an exchange of genetic material, and the "new" daughter cells are not the same as their parent's cells. Their genetic material has been changed by the conjugation ("sexual") process. The post-sexual cycle cells are like a newborn baby, they are composed of new genetic material. Again, the species is saved, but the individual is doomed even in simple eukaryotic single-celled species. It is not really the same individual after conjugation. We again have to return to the principal conclusion—aging is built into the basic biology of all living cells more complicated than prokaryotes such as bacteria and archaea.

To even begin to understand the process of cellular aging, we must appreciate the cause of the Hayflick limit, which prevents continuous cellular division in most eukaryotes on a molecular level. At least one answer may be telomeres.

Telomeres/Telomerase

Telomeres are DNA-protein structures found at the end of all eukaryotic linear chromosomes. One of the main characteristics of all eukaryotic cells is the presence of a nucleus containing the DNA. The DNA is separated into different double stranded linear structures called chromosomes. This arrangement is quite different from prokaryotic cells such as bacteria. In bacteria the DNA is not separated in a nuclear envelope. In prokaryotes, the DNA consists of a circular strand with **no free ends**. Chromosomes in eukaryotes have free ends, which raises the problem of how does the repair mechanisms for damaged DNA repair differentiate between a broken strand of DNA that needs repair and the end of the chromosome? Telomeres appear to be the answer. Chromosomes that lack end telomeres act like broken chromosomal ends and tend to fuse with each other and are subject to degradation by exonucleases (enzymes that chew up damaged DNA). Telomeres in eukaryotic cells act as a "cap" on the ends of the chromosomes to protect them. However, they do more than that as well. Human telomeres consist of double-stranded DNA a few thousand base pairs in length. The strand in Humans consists of a repeating sequence—$[GGGTTA]_n$, with one end extending beyond the double strand forming an "overhang." (Note: the letters represent the purine and pyrimidine bases that make up the DNA—G stands for guanine, T stands for thymidine and A stands for adenosine; see chapter 1-3, *Complex Chemistry to Life*). The subscript "$_n$" indicates that the sequence is repeated n times. The sequence varies in other eukaryotes. For example, the telomeres of Tetrahymena Thermophila, a single-celled ciliate protozoa that can be found in ponds, especially around decaying vegetation, has the sequence $[GGGGTT]_n$ in its telomeres.

When a cell divides there is a shortening in the length of the telomeres. The telomeres act as something of a replication clock. However, when the telomeres become short, the cell undergoes replicative senescence and stops dividing…they die of "old age." Telomeres are highly conserved, and they can be found on the ends of the chromosomes of all eukaryotic cells. They are what sets the Hayflick limit and appear to be responsible for replicative senescence. Although there are other important factors in aging (some of which we will discuss below), telomere shortening with each cell division sets the limit on the life expectancy of all Eukaryotic cells—from a single-celled protozoa growing in a pond to man. It is the molecular biology (chemistry if you will) of the telomeres that set your maximum life expectancy. The "telomere problem" appears to be an integral part of the **Methuselah Problem**. Fortunately, during a sexual cycle the "cellular clock" is reset to zero (full length telomeres). Otherwise, each generation (newborn baby) would have a very short lifespan.

How are the telomere set back to "zero" (full length). The answer is an enzyme that can add bases (like $[GGGTTA]_n$ back onto the telomeres, increasing their length. The enzyme is called (appropriately) telomerase. The "-ase" ending indicates that it is an enzyme. As you will recall (chapter 1-3), enzymes facilitate chemical reactions. However, this particular enzyme is composed of both amino acids (protein) and RNA, which is similar to ribozymes (previously discussed). Telomerase lengthens the telomeres and "rejuvenates" the cell (at least in the one aspect of telomere length and the number of times the cell can divide, which is generally referred to as replicative aging rather than chronological aging).

Telomerase is not generally found (expressed) in most somatic (body) cells but are expressed in germ cells such as the egg cells (oocytes) and sperm. Telomerase expression in Humans is also found in early fetal development, and in continually dividing populations of cells such as the gut lining and antibody-producing immune cells. However, biology (as usual) gets more complex. Telomerase is also generally expressed in cancer cells. Cancer cells are also immortal…they can continue to divide forever without reaching the Hayflick limit. We now understand that one of the reasons that cancer cells are immortal is because they continue to express the telomerase enzyme that prevents the telomeres from shortening. Cancer cells have solved the **Methuselah Problem** at least in terms of replicative aging and the Hayflick limit. In addition, the loss of the telomeres can result in chromosomal instability and cancer.

There is a clinical (medical) syndrome called Werner's syndrome, which is a Human condition with apparent accelerated aging. People with Werner's syndrome look old long before they should. They have grey hair with loss of hair, skin wrinkling and thinning with loss of skin elasticity, and they develop many diseases characteristic of old age such as cataracts, osteoporosis (loss of bone density), and atherosclerosis (the cause of heart attacks and strokes) and cancer. Short telomeres can lead to chromosomal instability, which can result in cancer. (http://dermnetnz.org/systemic/progeria.html)

Many of Werner's sufferers die of cancer by middle age (often in their 40s). How much (if any) telomere shortening contributes to cancer in a "normal" elderly population is not known. One problem of a telomere theory of cancer is that the types of tumors that occur in Werner's syndrome (such as fibrosarcoma noted in 10% of Werner syndrome patients) are rare in the general population.

Werner's syndrome has been demonstrated to be due to a mutation of a helicase gene that is involved in the unwinding of DNA. However, because it does unwind in the 3-primed to 5-primed direction of the DNA, it is probably not involved in the processes of DNA replication and transcription (which occurs in the opposite direction...5-prime to 3-prime direction) but may be involved in DNA repair. (http://www.bio.davidson.edu/Courses/Molbio/MolStudents/spring2003/McCord/wrn.htm) It has also been implicated in the maintenance of telomeres. The mutated helicase associated with Werner's syndrome can result in premature shortening of the telomeres and premature aging.

However, unfortunately, the expression of telomerase does not appear to be the simple answer for all aspects of aging in Humans. In fact, it is also part of the problem of cancer. As noted, telomerase expression and cellular immortality with unlimited growth is one of the basic characteristics of cancers. For example, Hahn and associates at M.I.T. showed that maintenance of telomere length was essential for continued tumor cell proliferation. Tumor cell expression of a mutant telomerase (that did not function properly) resulted in reduced telomere length and tumor cell death. Expression of this mutant telomerase even eliminated the tumorigenicity of these cells in vivo. (Hahn, W.C., Stewart, S.A., et al. *Nat. Med.*, vol. 5, no. 10, Oct. 1999, pp. 1129–30.)

It has already been possible to turn telomerase "on" in normal Human cells in tissue culture (in vitro). Bodnar and associates at the Geron Corporation and other associates at the University of Texas Southwestern Medical Center used a vector to transfer the telomerase gene into two telomerase-negative normal (non-cancerous) Human cell types. Telomerase-negative cells exhibited telomere shortening and senescence. Telomerase-expressing cells had elongated telomeres and continued to divide vigorously, and in addition, did not have the character of cancer cells.

It appears that the expression of telomerase with maintenance of telomere length is a general necessity, but not sufficient process in the development of cancer. It also appears to be a necessary process in extending the life expectancy of eukaryotic cells. Whether telomere lengthening would be a sufficient process to solve the **Methuselah Problem** is not known. The smart money bet is that telomerase expression in all somatic cells in man would probably not be a complete solution to aging, and if not rigorously controlled, it might be a serious problem in terms of cancer development. In fact, telomere shortening and the limitation it places on cellular division may be one metazoan mechanism for controlling cell growth and cancer that developed very early in metazoans. Think of the problem of rigidly controlling the growth of 30 to 50 trillion cells for 70 years and preventing any of those cells from becoming "independent" in terms of cell division. It is not surprising that clusters of independently dividing cells (cancers) occur.

What has often been noted is that it is surprising that many more cancers do not occur. The rigid control of cell division is one of the most important developments leading to the evolution of large multicellular creatures. Without such rigid control of cell division there would be total chaos at the cellular level of the multicellular creatures—animals. Turning telomerase on in every cell in the Human body without rigorous control might be like having every cell take the first step toward cancer. However, as appealing as the concept is that telomere shortening plays a role in the control of cell growth and limitations on cancer development, this cannot be the total answer since telomere shortening is even seen in single-celled protozoa. There are obviously no cancers in protozoa. Therefore, "Mother Nature's" rationale for telomere shortening and cellular senescence appears to predate metazoan development (about 700 million to perhaps 1 billion years ago). There is clearly more to this story. The rationale for the telomere system may possibly be embedded in the basic chemistry of chromosomes. For example, we noted earlier that telomeres serve as "caps" on the ends of chromosomes, so they do not look like broken strands of DNA to the DNA repair enzymes. Later in metazoan development, the mechanism may have been enlisted to also help control cell division and cancer.

There does appear to be a few possible exceptions to telomere shortening and continued growth. Generally, telomerase (which keeps the telomeres long) is expressed during embryonic and juvenile growth phases of most animals, but growth does not continue after adulthood and telomerase expression is not present in most adult body cells (somatic cells) as noted above. However, there are exceptions. The lobster (Homarus americanus) grows throughout its life, and the occurrence of senescence is slow.

High telomerase activity has been detected in all of the lobster's organs. Klapper and colleagues from Kiel, Germany has suggested that this telomerase expression is a mechanism for maintaining long-term cell proliferation capacity. (Klapper, W., et al. *PubMed*, 9849895.) This also appears to be true of the trout and mouse. Telomerase expression and continued growth appears to be the lobster's answer to aging. Unfortunately, the lobster's "strategy" of continued expression of telomerase throughout life to avoid aging is still questionable. No one has seen a thousand-year-old lobster the size of a school bus. Or at least they have not seen one and lived to tell about the experience, suggesting that there are other factors (and probably many other factors) other than continued telomerase expression involved in aging. However, continued and "controlled" growth with expression of telomerase might be one approach to aging in Humans where telomeres are only expressed in a few cells—the lobster solution. Hopefully with some modification of telomere expression, we could have the continued cellular growth and absence of aging without having to look like a lobster.

There are even some eukaryotic (single cell) exceptions to telomerase expression. A single-celled ciliate, Tetrahymena thermophili, and the yeast, Saccharomyces cerevisiae, appear to be immortal in terms of replicative aging.

However, there are other recent suggestions that the telomere/telomerase system is not the whole story in cellular senescence. Although cellular senescence (replicative aging) has generally been equated to cellular aging, this may not be the entire story. Senescent cells stop dividing or only divide sparingly, but they also have a decrease in cellular function, and sometimes even a change in basic

cellular function. For example, fibroblasts are an important cellular component of connective tissue. They lay down collagen, which is a major component of the intracellular connective tissue matrix—for example, beneath the skin. However, senescent fibroblasts secrete collagenase, which is an enzyme that breaks down collagen. (Campisi, Judith.
http://research.mednet.ucla.edu/pmts/sens/transbody.htm)

In other words, "old fibroblasts" are not only not doing their job of depositing collagen, but they are also actually breaking it down.

To make the picture of what is happening even fuzzier, there are even suggestions that sometimes senescence can occur without telomere shortening. One interesting observation is that senescence is dominant in cell fusion studies. When a senescent cell is fused with a young cell the resulting fused cell behaves like a senescent cell. (Clark, W.R. "Reflections on an Unsolved Problem of Biology: Evolution of Senescence and Death.") This suggests that the young cell cannot provide some "missing factor(s)" lacking in the senescent cell to make it "young." Instead, the senescent cells produce something that "ages" the young cell. This is the opposite result noted when most cancer cells are fused with a similar normal cell. Cancer-normal cell fusion experiments generally result in the cells behaving in a benign (normal or non-cancerous) manner.

There are even reports of gene mutations that prevent senescence regardless of telomere length. For example, a "knockout mutation" (where the gene product is no longer expressed—the gene is "knocked out") in the gene coding for P53 in mice results in the absence of cellular senescence. In essence, the cells do not grow old. This sounds like a "gimme putt" in golf (a sure thing) answer to the **Methuselah Problem**. Knockout gene expression of P53, and cells no longer senesce, and the **Methuselah Problem** is solved. Unfortunately, p53, the protein product of the P53 gene, is important in the control of cell division, and its absence or abnormal function may result in a tendency toward cancer. It is considered an oncogene...a gene involved in cancer formation (by its absence of function). Many cancers have a mutation in P53, which allows them to avoid senescence, and to continue to grow. Some viruses, such as the papovaviral implicated in cervical cancer, suppress P53 (and other genes), and start the cells on the road to cancer. In mice that have the knockout mutant P53, they do not undergo senescence, but they also do not live longer. They die early from a cancer or often from multiple cancers. There is also evidence that shortened telomeres activate the P53 pathway, which shuts down cell division. In other words, p53 activity can stop a cancer cell by making it undergo rapid senescence. (http://www.senescence.info/telomeres.html) This illustrates again the complexity of the basic biology of aging, and how intimately aging is associated with the problem of uncontrolled cellular proliferation—which is known as cancer. (Judith Campisi; http://research.mednet.ucla.edu/pmts/sens/transbody.htm)

However, regardless of the original biological reason for development of the "telomere/telomerase system," the ultimate solution to the aging problem may well involve a process for telomere lengthening, prevention of shortening or a mechanism to bypass the system entirely in terms of continued cell division, and to prevent the loss of cellular function associated with senescence. A solution to the telomere length problem will probably be a necessary, but not sufficient solution to the overall problem of aging. In other words, we probably cannot solve the **Methuselah Problem** without solving the telomere length problem, but solving the telomere length problem will probably not be adequate by itself to solve the **Methuselah Problem**.

There is a lot of potential in possible "telomerase therapy," and it might even be possible to wake up tomorrow and find a telomerase pill on the market that will lengthen your telomeres and solve the replication problem of cellular aging. However, again, that does not appear to be the ultimate solution to the chronological problem of aging, even though it certainly appears to be a part of the pie. The telomere/telomerase system appears to be an important player, but not the whole answer. The more immediate short-term pragmatic application of the telomere/telomerase system may be in the treatment of cancer. There are already telomerase inhibitors available, and under investigation for cancer therapy.

There also appears to be other very basic biological processes that may be fundamental in the aging process such as oxidative damage and DNA mutations and even senescent signaling pathways.

In addition, we will be discussing cloning, stem cells and embryonic stem cells shortly as a "treatment" for aging or a mechanism for "rejuvenation," but it should be noted that these cells have solved the telomere length problem because they have, in essence, gone through a sexual cycle, and have had their telomere length restored to that of a newborn baby.

Oxidative Damage/Mutations

We have previously noted that oxygen is a very toxic molecule (chapter 1-4: *Early Life to Complex Chemistry*). Oxidants can cause mutations in DNA, messenger RNA, and RNAs which have been increasingly implicated in regulating gene expression as well as damaged proteins. Over the years mutations, particularly in the DNA, can result in altered cellular function. There are many ways that DNA can be altered or mutated, including abstract things (at least in terms of daily awareness) such as cosmic rays or even radiation from rocks used to build our houses. However, oxidative damage from reactive oxygen species resulting from our own metabolism is also an important contributor to this process—perhaps the most important component. There are cellular mechanisms to repair mutational damage to DNA, and additional cellular mechanisms for reducing the level of reactive oxygen species—molecules that cause oxidative damage and even cause mutations in DNA. In fact, Ettore Bergamini from the University of Pisa estimates that there are one hundred million free radicals generated in our bodies every day, and ten thousand DNA lesions per cell per day. (Adams, J. U., quoting Bergamini, Ettore. "Autophagy & Longevity." *The Scientists*, http://www.the-scientists.com/2005/5/9/22/1/) That is a considerable amount of damage to sustain day after day. Perhaps the question should be "why do we not age even faster?" However, most animals, including man, have developed defense mechanisms to deal with free radicals and even to repair damaged DNA.

The defense mechanisms are very old and appear to be highly conserved. Highly conserved means that they are very similar in structure and function to enzymes found in lifeforms that evolved through separate evolutionary pathways—in some cases over 600 million years ago. Superoxide dismutase (a mouthful of chemistry which is often call SOD) and catalase are just two examples of proteins found in cells that help break down reactive oxygen species that damage proteins and cause mutations. SOD has similar morphology and function in yeast, the worm (C. elegans) and other vertebrates. SOD has 50% homology between yeast and mammals, including Humans. In fact, similar antioxidant repair mechanisms are even seen in prokaryotes (bacteria). These antioxidant enzymes appear to have evolved in response to the oxygen crisis (as we have previously discussed (chapter 1-4, *Early Life to Complex Chemistry*) that resulted from development of oxidative photosynthesis by cyanobacteria with significant accumulation of atmospheric oxygen between 2.5 and 1.8 billion years ago. Oxidative cellular damage is probably at least one important component of aging. Oxidative damage is particularly important in the energy-producing component of all eukaryotic cells—the mitochondria.

Mitochondria use oxygen to generate energy from substrates such as sugar (glucose). Unfortunately, despite an elaborate array of antioxidant defenses (we have only mentioned a few—SOD, Catalase), mitochondria are not completely efficient, and they continually produce "reactive oxygen species"—toxic oxygen-containing molecules. The resulting local damage to the mitochondria can be extensive. In fact, mitochondria are the major cellular source of oxygen free radicals. DNA damage has been noted to be 17 times greater in mitochondria compared to nuclear DNA damage. (Wallace, D. C., et al, *Curr. Genet.* vol. 12, pp. 81-90.) An increase in DNA damage and mutations can adversely affect mitochondrial function. In turn, as the powerhouse of the cell, decreased or altered mitochondrial performance may severely affect normal cellular metabolism and function. It appears that at least one component of aging is a direct consequence of the fact that eukaryotic cells "ate" (phagocytosed) a bacteria billions of years ago that became a mitochondrion and utilized the bacteria's mechanism of oxidative metabolism as an energy source. Clearly eukaryotic cells gained considerably in terms of metabolism, but it now appears that we pay a price in terms of cellular aging.

This decrease in mitochondrial function appears to be an independent age-associated dysfunction. However, mutants in both fruit flies and mice which produce large excesses of superoxide dismutase (SOD) age just as rapidly as other animals (*Proc. Natl. Acad. Sci.,* vol. 87, 1990, p. 4270.) Therefore, even though oxidative damage of DNA (and especially mitochondria DNA) appears to be an important component of the aging process, increasing enzymes that limit such oxidative damage does not appear to be an "answer" to the problem (or again, at least not the whole answer).

In other experiments, genetically engineered mice that carry a mutated enzyme called DNA polymerase-gamma, which normally "proofreads" DNA and aids in DNA repair in Mitochondria have three to five time the number of mitochondrial DNA errors as normal mice and showed early signs of aging and reduced lifespans. None of the mutated mice lived more than 60 weeks compared to normal mice that generally live about 100 weeks. (Trifunovic, A., et al. *Nature*, vol. 429, 27 May 2004, pp. 417–423). This again demonstrates how an increase in mutations in the mitochondria can result in early aging and a decrease in life expectancy, however, this does not necessarily imply that a decrease in mutations would result in living longer, and particularly living for a thousand years, an answer to the Methuselah problem.

Antioxidants

There has been some progress in research involving prevention of oxidative mitochondrial damage. Bruce Ames and colleagues in a series of articles in the *Proceedings of the National Academy of Science* (probably the most prestigious scientific journal in the world) have reported significant improvement in the memory of older rats, and better function of several parameters of mitochondrial function. (Hagen, T.M. *PNAC,* vol. 99, no. 4, Feb. 19, 2002.) Ames and his colleagues used alpha-lipoic acid (ALA) and acetyl-L-carnitine (ALCAR) in combination. The enzyme carnitine acetyltransferase is also important in mitochondrial metabolism, and its function decreases with aging. This results in poor mitochondrial function. The acetyl-L-carnitine is the substrate for this enzyme. By providing more substrate, the enzyme and therefore the mitochondria function better. However, the addition of the ALCAR also resulted in increased reactive oxygen species production which could further damage the mitochondria and the cell. The addition of lipoic acid decreased or neutralized the increased oxidant production. Although these studies have demonstrated some interesting biochemical and functional improvements in old animals, there is no evidence currently that the animals live longer, however studies of lifespan are currently being done.

Vitamins are the Answer?

Another strategy to help prevent oxidative damage is the use of vitamins, especially C and E. Remember Linus Pauling who was the "radical" chemistry professor from California Institute of Technology who was opposed to nuclear testing? He was also an early advocate of high dose vitamin C intake. Pauling had noticed that 25 million years ago the ancestor of hominids lost the enzyme necessary to synthesize ascorbic acid (vitamin C). All mammals except Humans and some other primates, guinea pigs and a fruit-eating bat make their own vitamin C endogenously (they biochemically synthesize it themselves, and do not need an external source).

A lack of vitamin C results in the disease scurvy. Scurvy was quite devastating to sailors on long voyages in the past. Vasco da Gama, a Portuguese navigator, sailed from Lisbon to Calicut, India in 1497/1498 with 160 sailors. Only sixty arrived at their destination. The others had died of scurvy during the journey. (Pauling, Linus. *Vitamin C the Common Cold and the Flu*. San Francisco: W.H. Freeman and Company, 1976) Even as late as 1740, the British Admiral George Anson set out with a squadron of six ships and 961 sailors. When he reached his destination there were only 335 sailors with more than half dying of scurvy. It was the British navy (actually a Scottish physician James Lind in the British navy in 1747) who recognized that they could prevent scurvy on long ocean voyages by using lemons (high in vitamin C), and that is why British sailors are sometimes referred to as "limeys." The recommended Human daily allowance of vitamin C is only 60 milligrams a day, which is enough to prevent scurvy.

However, Pauling noted that animals that make vitamin C produce about 10 grams (10,000 milligrams) per day scaled to a 70-kilogram body mass (like a Human). Pauling recognized its benefit in the common cold, influenza, and many other conditions, including its use as an antioxidant. However, this is still very controversial in the medical community. Pauling began to recommend "massive" doses of vitamin C—as high as 10 to 16 grams a day. Critics have charged that such high doses only benefit the pharmaceutical companies that manufacture vitamin C, and simply result in expensive pee (urine). However, Pauling lived to be 93, while following his own advice and using "massive" doses of vitamin C, but the megavitamin approach to a long life is still very questionable.

For example, high doses of fat-soluble vitamins can be toxic. Yet, there are an incredible number of people who "keep the megavitamin (and herbs) faith." I used to consider them to be just another "nut group."

My first encounter with the vitamin faithful was not a good one. I was called to the emergency room (during my residency training) to see a man who had been involved in a motorcycle accident. Unfortunately, he had used his head as a "brake" on a concrete highway. In addition to scraping off the skin of his scalp he was unconscious. The neurosurgeons were there looking at his squash, and the orthopedists were looking at his fractured leg and arm. However, an even more pressing problem arose when he began to drop his blood pressure, and we discovered he had lots of free blood in his abdominal cavity, and he was going into shock (this is cardiovascular shock, not what the layman generally regards as shock). He was bleeding into his abdominal cavity! Everyone rapidly agreed that general surgery would have the first shot at him to control his internal bleeding (otherwise there would be nothing left to treat). Then the other teams could do "their thing." General surgery meant the chief resident and myself (I was doing a rotation on general surgery at the time). We rushed him to the operating room, removed his ruptured spleen and repaired some liver lacerations, and gave him some fluid and blood transfusions to stabilize him. After some tests, the neurosurgeons decided that he did not have an intracranial bleed, and they treated him medically to try and reduce the swelling of his brain. The orthopedist reduced his fractures and placed his arm and leg in a cast. It had been a long night, but by morning we had gotten him into the surgical intensive care unit…alive. Over the next few days, he stabilized, but he was still unconscious, and he was on a ventilator.

The man's wife was visiting, and asked the chief resident how her husband was doing. I remember the chief resident trying to be both optimistic and realistic, but he concluded that her husband was in a very serious condition. A few days later, we had an almost identical conversation, but after hearing that there were no changes in her husband's condition, the wife asked, "if we were giving him vitamins daily." The chief resident turned to me. For those of you not familiar with large teaching hospitals, the chief resident was about as close as you could get to God in the hospital, and they rarely took care of such "minor details" as intravenous fluid orders, or monitoring urine output, or changing bandages, or checking electrolytes, or checking the hemoglobin and hematocrit…those were the job of someone lower on the totem pole…the second-year resident (myself in this case). I had to inform the wife that we were not giving him vitamins. She then pulled out a book written by a megavitamin profit of the 1960s (I will refrain from giving the author's name) and began to explain to us (the team of residents taking care of the the man) "that" was the reason her husband was not getting better. After listening patiently for a few minutes, the chief resident turned to me and instructed me to give him vitamins. I went to the pharmacy, and the man got vitamins…every day in his IV fluids. A few days later, we again had an almost identical conversation with the man's wife.

"He was still in a very serious condition!" In fact, he was becoming worse. We had to give him increasing concentrations of oxygen by the ventilator to maintain his blood oxygen concentration. That was definitely not a good sign.

"Yes, we were giving him vitamins every day."

However, on hearing that her husband was not getting better (again), she wanted to know if we were giving him vitamin E. Frankly, I did not know. The only thing that I remember knowing about vitamin E at the time (1960s) was that it was reported to be a sexual stimulant for rats (which still holds today, but now other functions of vitamin E are much better understood). I had to go back to the pharmacy and look it up. Unfortunately, I had to report back…no vitamin E in the multivitamins. Out came the book again. The chief resident recognized that we were not treating the man, we were treating the woman and her grief. I had already been told by the chief resident to give the patient anything reasonable that the wife wanted provided that I did not kill the patient. She showed me right in the book the number of units of vitamin E that he should be given every day (I wrote it down). There was an immediate problem; the pharmacy did not have any vitamin E. They had never had a request for vitamin E. The pharmacists asked:

"What was it needed for?"

"Trauma from a motorcycle accident."

That definitely did not impress the pharmacists. As a second-year resident, it did not look like I was going to get any vitamin E.

To get a new "drug" stocked by the pharmacy, I was told that I would have to go to the pharmacy committee and argue my case. In addition, there was the dosing problem. It was not clear that we could get that much vitamin E into the man (in fact, it wasn't clear, whether there was that much vitamin E anywhere in the state of Florida or possibly even in the entire U.S. for that matter at that time). The man died shortly afterwards. He had never gotten the vitamin E. This was my first encounter with the megavitamin religion. It is justifiable to call it a religion because it shares the common factors of believing something based on hope and justified by faith. Unfortunately, vitamins and herbs and other "quick fixes" are a common final pathway for many "futurists" who recognize that we are very near the critical point of controlling our own evolution and extending our lifespan. Regrettably, this is nothing more than the same old "religious answer" of Heaven wrapped in a vitamin pill. The real answer lies in the continued research that may produce the data and the genes or approaches that will be needed to make significant progress in extending mankind's life expectancy. Megavitamins and herbs do not appear to be the answer to a long life. However, in the past the lack of vitamin "c" for sailors on long voyages on ships definitely paid a high price for not having adequate vitamin "c" intake.

It appears reasonably clear currently that vitamin C is a valuable antioxidant. Vitamin C is also probably of value in the prevention of some infectious diseases and may also be of value in other conditions as well. However, although antioxidants may decrease the oxidative damage of cells, there are no studies that have demonstrated that anyone will live longer taking megavitamins including vitamin "C". In fact, the "data" appears to be exactly the opposite. Bjelakovic and associates reviewed 395 publications involving 232,606 patients using vitamin A, vitamin E, beta carotene, vitamin C and selenium. They reported a significantly increased mortality in patients using vitamin A, vitamin E, or beta carotene. (Bjelakovic, G., et al., *JAMA*, vol. 297, 2007, pp. 842–857.) Vitamin C and selenium had no effect on mortality (provided there was an adequate intake). Likewise, in another study looking at the use of a statins (a class of drugs used to lower low density lipoproteins, which have been implicated in coronary artery disease) compared to a placebo, antioxidant vitamins (A, E, C) or vitamins combined with a statin, they found an increased incidence of adverse cardiac events in the placebo group, and also in the vitamin group, and even in the statin combined with vitamin group. Other studies on the progression of atherosclerotic plaques in the coronary arteries showed a decrease in progression using a statin, but an increase in progression in the placebo group, and in the vitamin group and even some progression in the vitamin plus statin group. This implies that the antioxidant vitamins were hindering some of the positive effects of the statin. (Brown, B., et al. *Atherosclerosis, Thrombosis, and Vascular Biology*, vol. 22, 2002, p. 1535.) Whether this is confirmed, time will tell. Considering that atherosclerosis is currently the leading killer, and the main reason that you and I may not live to be 100 years old, these findings are very significant. I no longer take vitamin E. It has been demonstrated to be an excellent antioxidant, but it appears to be involved in other ways that are detrimental.

We should recall that mice and most other animals synthesize relatively high doses of vitamin C, and yet the lifespan of a mouse is only 2 to 3 years. Although oxidative damage and mutations from oxidants, and especially mitochondrial oxidative damage, is one of the leading theories regarding aging, it is not clear at this point whether this is the primary mechanism in aging or only one of many players, and there is no good evidence that correcting oxidative damage by current techniques has made any significant changes in life expectancy. Vitamin C and other antioxidants are probably helpful in reducing oxidative damage, but they are clearly not the answer to the **Methuselah Problem** for mankind.

Apoptosis

Apoptosis is a process of programmed cell death—cells on command literally commit suicide. This is quite different from cells dying from injury. Apoptosis is a process of cells dying by self-destruction; in essence, cells following a "program" leading to death after receiving a "signal." The death signal may come from within the cell or from outside of the cell. One (of many) examples of an outside apoptosis signal are the **Fas/FasL** "system." **Fas** and **FasL** act through a lock and key mechanism. **Fas** may be found as a receptor on the outer membrane of many cells. **FasL** may be found on the outer membrane of cytotoxic T lymphocytes—cells that patrol the body looking for abnormal cells. T lymphocytes are part of the immune system. When a cytotoxic T lymphocyte detects an appropriate abnormal cell—for example a virus infected cell or perhaps a cancer cell—**FasL** (on the T lymphocyte) binds to the **Fas** on the target cell. By this binding, the target cell is "signaled" to undergo apoptosis and die.

Internal cellular signals resulting from cellular damage from reactive oxygen species or other types of damage can also trigger apoptosis. The cell recognizes that it is damaged and initiates its own internal signal for apoptosis. Again, several internal death-signaling mechanisms have been described.

In the actual apoptosis process the cell literally digests its own DNA and other important macromolecules (such as proteins).

Surprisingly, apoptosis is a very basic cellular process that can be traced back to the origin of the Metazoan (multicellular creatures) with roots even in single-celled eukaryotes.

Programmed cell death or apoptosis is necessary even for proper embryonic development. For example, during embryonic development there is a web of skin between the fingers and toes in Humans. This web of skin is removed by cell apoptosis. The arrangement of the proper connections among neurons in the brain requires some cells to be eliminated by apoptosis. Apoptosis is common in all metazoans. For example, the single-celled nematode, C. elegans, forms 1090 cells during development, however, exactly 131 of those cells are eliminated by apoptosis. In fact, using genetic engineering, it has been possible to place a Human gene into the worm, C. elegans that can prevent or reduce one type of

programmed cell death, showing how this mechanism has been conserved over 570 million years since mammals last separated from worms. (Vaux, D.L. *Science*, vol. 258, 1992, pp. 1955–1957.)
(http://www.informatics.jax.org/silver/1.3.shtml)

In fact, laboratory experiments have even shown that some mammalian genes that trigger apoptosis can also result in apoptosis when transferred to yeast, and genes that block apoptosis in mammals can also block apoptosis when transferred to yeast. Should I stop mentioning evolution? Here we have a basic biologic process that is very similar in yeast to a worm to Humans. In fact, so similar that a gene that works in mammals can also do the same "job" in yeast.

In addition to telomeres and the Hayflick limit to cell division, and oxidative cellular damage and mutations, and even very conserved cellular processes like apoptosis, there are other components to aging.

Genetics

Genetics undoubtedly plays an important part in aging. There have been more than 200 genes that have been linked to Human aging in one form or other, and there will probably be more found. (www.genomics.senescence.info) We have talked about alleles before; they are the changes or variations found in genes. Very little is known about the allelic variations in the 200 or so genes suggested to be associated with aging. However, I will "stick my neck out," and predict that the "best set" of alleles in genes associated with aging (those that together will give the longest lifespan) will be well-known by the end of the 21st century (and possibly a lot sooner), and having that set (we can call it the LLAS or Long Life Allelic Set…it will probably be a subset of the Deus Allelic Set (a proposed set of genes of a future God-like creature) will push the lifespan of mankind into the range of the longest living known creature—the giant tortoise (Testudo Elephantopus) with a reported lifespan of up to 188 years as noted previously. Notice that this prediction requires no new Human genes, just the right set of alleles already present. Each of us with long-lived parents probably has at least a few of the right alleles. However, the strategy is to get all of them.

In addition, we might find a few genes in the giant tortoise that would be of value, and perhaps the lobster's solution to telomere length would be of benefit (hopefully without increasing the incidence of cancer).

How would we get all these genes into our genome? Artificial Human chromosomes have already been synthesized and inserted into Human cells and have been passed on following cell division…the 47th Chromosome. Placing the right alleles and genes in such a chromosome is really a technical problem that is within the reach of current genetic engineering methodologies. Most likely this will ultimately all be a part of the **Optimal Human Allelic Set(s)** and eventually the Deus Allelic Set, when the whole Human genome is finally genetically re-engineered.

However, we are nowhere near a genetic answer to aging at this time. We share 98.4% homology in our genes with chimpanzees, yet we live twice as long. (Coles, L.S. http://www.grg.org/resources/blindmen.html) Clearly, the chimp has the wrong set of genes or alleles for a long life. In fact, the considerable variation in lifespan of different species is one of the strongest indications that genes may play a significant role in a long life. Even in Humans, the best formula for a long life is to have parents who lived a long life. (Mill Hill Essays. http://www.nimr.mrc.ac.uk/MillHillEssays/2001/ageing.htm)

Longevity Genes

However, let's look at some specific genes. Several "clock genes" have been identified in the nematode worm (C. elegans) that regulate overall cellular metabolism (energy production and the generation of small molecules needed for life) and have a major effect on lifespan. Variants of one gene called **Age-1** can result in nearly a doubling of the lifespan. Specific gene variants (alleles in the jargon of genetics as we have discussed) that are associated with a long lifespan are what we would expect from our discussion of aging. They are associated with decreased DNA damage, increased ability to repair DNA damage, decreased mitochondrial damage, and increased intracellular antioxidant enzymes such as SOD and catalase.

Unfortunately, a great deal of the research on aging and the genetics of aging has been done in short-lived creatures such as the fruit fly with a life expectancy of only 21 days or the nematode worm (C. elegans) with a life expectancy of 20 days or the mouse with a life expectancy of 2 to 3 years. In all these animals, studies have been done that have significantly increased their lifespan. However, it is not clear whether such techniques or genetic manipulations are directly applicable to man (except perhaps calorie restriction to be discussed below). There have been repeated criticisms even from scientists studying aging as to how much a short-lived species can really teach us about extending the lifespan of "long-living" creatures such as Homo sapiens. However, it was discovered that yeast has an ortholog (a similar gene sequence found in different species and derived from a common ancestor generally with similar functions) of a Human gene in which mutations cause Werner's syndrome (discussed above). Another possible important exception is a gene called Silent Information Regulator or SIR2 found in yeast, C. elegans, and appears to be universally conserved from yeast to mammals including man (there are seven mammalian homologs called SIRT1 to SIRT7). In both yeast and C. elegans, adding extra copies of the SIR2 gene extends their lifespan. Conversely, deleting the SIR2 gene shortens their lifespan. SIRT1 (the mammalian version of SIR2) increases survival of mammalian cells in culture. SIR2 has also been suggested as the molecular mechanism for extending lifespan by caloric restriction. More recently, multiple members of the SIR2 family (not just SIR2) have been implicated in regulating lifespan in higher organisms. (Lamming, D. W., et al. *Science*, vol. 309, 16 Sept. 2005, p. 1861.) In addition, the mechanism of action is turning out to be complex with many other players, at least in yeast. In fact, in yeast, the SIR2 response to caloric restriction appears to act through a mechanism not found in vertebrates (which again raises the argument of how useful findings in yeast in are studying mechanisms of aging in vertebrates), and in addition, there are alternate pathways (other than SIR2) that are activated in response to caloric restriction that result in extending the lifespan of yeast. (Kaeberlein, M., et al. Science, vol. 310, 18 Nov. 2005, p. 1193.)

However, SIRT1 also interacts and down-regulates p53—the tumor-suppressor protein. Downregulating p53 could result in an increased risk for cancer. (Guarente, Leonard P.
http://www.hms.harvard.edu/armenise/old_site/4b_link.htm
and http://www.globaltechnoscan.com/29march-4thapril/gene.htm)

This appears to be an interesting line of research, but a great deal more needs to be learned about the function of the system before any realistic application could be applied to aging in Humans. In the short haul, this is one of the most interesting areas of research. Leonard Guarente, who's lab at MIT has been working on aging and the SIR2 gene for over a dozen years feels that the increase in life expectancy by perturbation of the SIR2 gene homologs in Humans is a very realistic near-term possibility, however, he also predicts that the potential for life extension would be modest—perhaps on the order of a decade. Although any extension would be a great scientific breakthrough, if for no other reason than the proof of principle (aging can be delayed and lifespan extended in Humans by manipulation of genes). However, it also appears that the SIR2 gene and its mammalian homologs are not the answer to the **Methuselah Problem**.

Another gene of interest is called Klotho. It was originally identified as a mutated gene in mice with accelerated development of age-related disorders like those noted in Humans. Mice homozygous (both gene affected) for the mutated gene (which is non-functional) usually die around 2 months of age (compared to 2 to 3 years for normal mice). Klotho appears to function as a circulating hormone that represses intracellular signals of insulin and insulin-like growth factor 1 (IGF1). Klotho appears to be an evolutionarily conserved mechanism for extending lifespan. Even some single-nucleotide polymorphisms (an alteration in only one base pair in the gene) in the Human Klotho gene are associated with altered lifespans and altered risks of coronary artery disease (associated with heart attacks and strokes), and osteoporosis (bone loss associated with fractures). Klotho overexpression has also been noted to extend the lifespan of mice 20 to 31% (in males depending on the particular Klotho allele being expressed). (Kurosu, H., et al. *Science*, vol. 309, 16 Sept. 2005, p. 1829.) The mechanism of action appears to be independent of caloric restriction.

Probably the best example of genetically "programmed" aging is the Pacific salmon. Pacific salmon age incredibly fast, and they die shortly after spawning. There is striking somatic and neural degeneration due to a marked elevation in plasma cortisol levels (a hormone produced in the Human adrenal gland often referred to as a "stress" hormone) resulting in sudden and dramatic catastrophic senescence and death. Although the Pacific salmon is an example of programmed aging, it is not clear that this type of aging has any relevance to Humans other than in the general sense that all aging may have a genetically programmed component—although not as rapid and impressive as the Pacific salmon.

Hansen et al. from the University of California at San Francisco (Hansen, M., et al. *pLoS*, vol. 1, no. 1, July 2005.) reported on 23 new longevity genes that appear to work through at least three distinct regulatory systems. Again, this is in C. elegans, and it is not clear how much of this can be translated into Humans. However, it does reflect how fast this field is moving, and suggests that in addition to oxidative damage, mitochondrial damage, DNA damage and repair mechanisms, and telomeres, there appears to be a set of genes that are specifically involved in regulating aging through several sets of signaling pathways (including dietary restrictions). The hope is that perturbation of these pathways (with drugs or RNA interference, which was used in the Hansen et. al. study) may be on the fast road to real results in aging…soon.

Stem Cells and Regeneration

The body of any higher multicellular organism is composed of specialized cells that perform functions, such as muscle cells, nerve cells, blood cells, liver cells, skin cells, etc. This specialization of function is called differentiation. A differentiated cell cannot change its function. Once a cell has become committed to a particular function by differentiation, it usually remains that cell type—such as a muscle cell or skin cell, etc. A muscle cell placed in the brain remains a muscle cell or a nerve cell placed in the liver does not function like a liver cell. Some cells, such as skin cells or the cells lining the gastrointestinal tract, continue to divide throughout life, while other cells, such as muscle or nerve cells, do not divide in an adult. In other words, different cells have a very different approach to long-term survival. Continually dividing cells will ultimately reach the Hayflick limit, and undergo senescence, and no longer divide, while non-dividing cell populations will be gradually "lost" from a variety of external and internal factors but will not be replaced.

However, all cells still retain the complete set of genes or DNA codes as every other cell—differentiated or not. Although the details of the genetics and molecular biology of differentiation are still being worked out, it appears that the major difference between different differentiated cells, such as muscle or skin, involves which sets of genes are turned on (expressed), and which genes are turned off. This concept of cell differentiation had until very recently been "set in stone"— "once a skin cell, always a skin cell."

However, there have been two recent developments that have clearly demonstrated that "It ain't necessarily so!" The ability to find or produce undifferentiated cells to act as a "replacement part" or at least an understanding of the genetic and biochemical processes involved may ultimately have a profound effect on the treatment of some disease processes and possibly on aging as noted below.

One development was the birth of Dolly, the cloned sheep. Dolly was the product of research done by Ian Wilmut, an embryologist at the Roslin Institute in Scotland. The nucleus used to produce Dolly was taken from a gland cell (a differentiated cell) from the breast of a Finn Dorset ewe. The nucleus of the gland cell was inserted into the egg cell of a Scottish Blackface sheep after its nucleus had been removed. The fertilized egg began to divide, and the embryo was placed within the uterus of a sheep. There was a normal gestation, and Dolly was born. Dolly has the white face of a Finn Dorset (the DNA donor) rather than the black face of a Scottish blackface sheep—the enucleated egg donor.

Although the technique has resulted in a number of scientific "breakthroughs," including the realization that animals can be produced from differentiated cells and be genetically identical to the nuclear donor, it is not a totally perfected technique. Dolly was the only success out of 277 reconstructed embryos, and she was euthanized at age 6 for medical reasons including arthritis and lung disease. Most sheep live to be 11 to 12 years of age. However, there have now been many more cloned species since Dolly, including additional cloned sheep. Some clones have also included genetic manipulations—genetic engineering. However, Dolly proved that fully differentiated cells can be reprogrammed. We will skip over the potential for cloning a fully developed Human, which is quite controversial, and would really add nothing new scientifically other than a "we did it" medal. This type of cloning has become known as reproductive cloning to distinguish (and separate) it from cloning used to produce cells that may be of value for medical treatment and/or the "treatment" of aging. The current experience indicates that the cloning technique does not appear to be fully perfected at this time as per the Dolly experience (and others) and should clearly not be undertaken for the sake of the cloned individual (child) until the technique is well understood and even provides some advantage for the cloned individual. However, most of the current criticisms regarding cloning center around the "moral" issue, which we have discussed. The fact is that a cloned baby (if and when the technique is fully developed and does not result in any detrimental medical problems for the cloned individual) is simply a different technique for producing a baby. Although a cloned baby would share the DNA with its single parent, it is not its parents, just as any other baby is not its parents. Babies share the genes of their parents, but as any parent knows—a baby is not the parent. Identical twins are "clones," they share identical DNA, and in addition, they often share similar (but never identical) environmental surroundings, however, they are different individuals. What is the moral outrage against twins that share identical DNA? Twin can be considered a clone produced by Mother Nature (God if you prefer). In fact, every baby is the half-clone of each of its parents. Yet, there is "moral indignity and outrage" that someone might be a clone. The question of nurture versus nature—the influence of the environment and experience versus the effects of genes—has been argued for years. Currently, advocates of both influences have recognized the important contribution of both components in the shaping of the "self" that creates an individual. Arguably, nurture and nature each contribute about half. The important point is that two identical twins may look alike (identical), but they are each individual.

The fact is, those that try to suppress research in areas of cloning, stem cells or embryonic stem cells based on God are morally confused and are terribly misguided. The usual reasoning that "I hear" is that removing the nucleus of the egg cell kills it, and that is a form of an abortion. However, putting a skin cell nucleus back into the egg cell has revived it. No matter, it is still a sin and immoral, because the original egg nucleus was removed. However, the egg had never been fertilized. No matter, it had the "potential" to become a Human, and therefore, removing the nucleus is a sin and immoral. Under this scenario, the egg cell appears to have taken on some form of "holiness." Let's consider another scenario. A young woman is a virgin. She goes into a convent to become a nun. Each month she ovulates, and each month the egg is not fertilized and dies. However, not only is this not a sin, but it is also "holy." Letting an egg cell die for God is not holey! But the egg is dead, blessed be God! The egg has intentionally been allowed to die. What happened to the "holiness" of the egg? Apparently, the egg is not "holy" after all, it is just a cell. The problem lies in trying to apply biblical ethics to situations where they were never intended to be applied. In fact, reasoning with basic **Human-Centered Ethics**, the critics of this type of research are immoral since they potentially harm Humans in the name of God or for some other "higher ethical principle." In **Human-Centered Ethics**, there should be no higher ethical principle than the "good" of the individual within the perspectives of the "good" of mankind.

Of course, reproductive cloning to produce a baby has nothing to do with aging, however, studies of the cloning process could have a great deal to do with aging. In cloning, the nucleus of a mature (differentiated or specialized) cell (such as a skin cell) is placed in an egg after the egg nucleus has been removed. Something within the cytoplasm of the egg reprograms the skin cell nucleus, which then develops the capability of again making every cell in a new individual. However, in addition to becoming an undifferentiated cell, the cloning process also resets the age of the "old" cell (the skin cell in our example) to that of a newborn. In other words, something has happened to the "old" cell during the cloning process. It is no longer an old cell—the Hayflick limit has been reset. What did the cytoplasm of the egg do to the old skin cell nucleus to make it "young" again? We do not know, but this is the only real example that I am aware of where an old cell is rejuvenated with all the potential of a cell of a newborn. Understanding exactly what is happening during the cloning process can shed considerable light on both cellular differentiation, embryogenesis, and aging. The other potentially very important use of cloning is in the generation of embryonic stem cells.

Embryonic Stem Cells

The second important finding regarding cell differentiation is the recognition of undifferentiated "reserve cells" called stem cells. Stem cells are undifferentiated (or minimally differentiated) cells that have retained the capacity to divide forever—at least in tissue culture. Stem cells retain the ability to differentiate into many different cell types (depending on the type of stem cell). A fertilized egg is the best example of a totipotent stem cell, which has the potential ability to give rise to all the different cell types in the body, such as brain, muscle, liver, heart, etc. After fertilization, the (now) embryo begins to divide, and continues to divide until a "ball" of cells are formed—the morula. These are all embryonic stem cells and have the potential to divide into any cell type in the body. However, after about four days of cell division the embryo begins to develop into an outer layer of cells that will form the placenta, and an inner cell mass that will develop into the fetus. The cells of the inner cell mass have "differentiated slightly" or become specialized. They still retain considerable potential for developing into all the cell types in the body, but they have at least differentiated or specialized to a small extent. These minimally differentiated cells are pluripotent stem cells. Initially it was thought that only these embryonic stem cells had the potential to differentiate into all the different cell types in the body.

However, recently bone marrow stem cells have been identified in adults that can be coaxed to differentiate into at least some different cell types (other than bone marrow). Whether these bone marrow or other stem cells have the potential to differentiate into all cell types or if they have the same potential for growth as embryonic stem cells is yet to be determined. The first adult stem cells were not identified until 1998. The "Dolly" experience would indicate that fully differentiated cells can be reprogrammed into germ cells. Techniques for identifying other types of stem cells in adults, and the potential for reprogramming these cells into almost any type of differentiated cell is being actively pursued. The use of such reprogrammed stem cells to replenish tissues has recently been referred to as "regeneration." Exactly what is the potential for tissue or organ regeneration using stem cells is not clear at this point. There is also considerable controversy as to whether embryonic stem cells (with what is thought to be an unlimited potential for cell division and are taken from early developing embryos) will be necessary or whether adult stem cells can be adequately reprogrammed for the job. In fact, the placement of a differentiated cell into an empty egg with subsequent cell division and development of embryonic stem cell (cloning) is nothing more than one technique for cellular reprogramming. Unfortunately, the same well-intended but misguided individuals have attacked such attempts at reprogramming cells as immoral and, of course... sinful. The President(s) and the Congress of the United States has gotten into the debate, and counted votes, and outlawed the development of any new lines of Human embryonic stem cells for research purposes. The potential importance of embryonic stem cells in the treatment of disease processes or as "regenerative cells" in aging is best illustrated by comparison with organ transplants.

Organ transplants of heart, kidneys, liver and even lungs have the potential to usher in an era of "replacement parts." However, the problem of tissue or organ rejection has severely limited the potential for organ transplantation both because of the continued need to prevent rejection (with drugs that suppress organ rejection, but also suppress the immune system in terms of proper function such as fighting infection or tumor surveillance), and the difficulty in finding adequately matched donors (better matched organs lead to less vigorous attempts at organ rejection). There is now a new potential therapy on the horizon for "transplantation" using stem cells. However, clinical Human experience now is extremely limited. The stem cell approach is an entirely new paradigm. Instead of transplanting whole organs, undifferentiated cells are infused into an organ to replace dead or injured cells—in essence to try to "rejuvenate" the organ with "young cells." The problem of tissue rejection is bypassed since the cells used are taken from the same individual that will again receive the cells back. In addition, the use of embryonic stem cells would give an almost limitless supply of cells, and again, they are the recipient's own cells so immunologic cell rejection should not be a problem.

Although the research using stem cells has just begun, there have been some preliminary reports showing the potential for stem cells. Donald Orlic at the National Human Genome Research Institute in Bethesda, Maryland noted that injecting bone marrow cells into the hearts of mice with an induced heart attack helped repair the damaged heart muscle. The bone marrow cells were rich in stem cells that were thought to have helped repair the heart damage. This line of research using bone marrow stem cells would have been expected to take years to ever reach Human clinical trials. However, the need and potential benefit was so great considering the large number of people suffering from heart attack who survive the heart attack, but who are left with a damaged and poorly functioning myocardium (heart muscle) that clinical trials were undertaken almost immediately.

In March 2001, Bodo-Eckehard Strauer and colleagues at the University of Dusseldorf in Germany, injected bone marrow cells into the heart of a 46-year-old heart attack survivor with over a third of his heart damaged. Ten weeks later, the damaged myocardial area had decreased by roughly a third. Strauer and colleagues have now completed 60 patients (as of April 2004) (Couzin, Jennifer. Vogel, Gretchen. "News Focus." *Science*, vol. 304, April 9, 2004, pp. 192–194.) In addition, Andreas Zeiher at the University of Frankfurt has treated 34 patients, and Helmut Drexler at the University of Hannover has treated another 30 patients. Preliminary results suggest that patients treated with stem cell therapy have recovered between 5 and 30% of their lost pumping capacity. There have not been any serious complications.

Researchers at the Pró-Cardíaco Hospital in Rio de Janeiro and the Texas Heart Institution in Houston, Texas have directly injected bone marrow stem cells into the hearts of 14 patients. The technique of stem cell injection involves making a small incision in the groin and threading a long catheter into the heart—a process called cardiac catheterization. Cardiac catheterization is routinely used to perform diagnostic studies of the coronary arteries (coronary arteriography) for atherosclerosis (hardening of the arteries that can cause "heart attacks"). However, in this case, they used a needle on the end of the catheter to inject millions of stem cells into separate areas of the myocardium (heart muscle). Many patients have noted marked improvement in cardiac function. These are patients with advanced myocardial disease, most of whom have already had the usual conventional therapies such as coronary bypass surgery and/or angioplasty of the coronary arteries.

One patient, Nelson Aguia had 11 areas of his heart injected with stem cells. Prior to the procedure, he had sustained extensive myocardial damage (heart attacks) and was no longer able to work or even walk any significant distance. Within six months of stem cell therapy, he was walking three miles a day, swimming laps in a pool, and had returned to work full time. Aquia has stated, "I was planning my funeral…" but now "I'm ready for anything…This treatment has given me life." However, it is not clear whether the stem cells are becoming new heart muscle cells or stimulating the growth of heart muscle cells. Nelson Aguia only cares that it worked for him. (http://www.tmc.edu/tmcnews/06_01_02/page_01.html)

Philippe Menasche' at the hospital European Georges Pompidou was the first to use cell-therapy in June of 2000. He injected thigh muscle cells (rather than bone marrow or stem cells) into the heart during bypass surgery. Four of the ten patients developed cardiac arrhythmias. What potential this type of cell-therapy would have for post myocardial infarct patients in the future is not known. (Couzin, Jennifer. Vogel, Gretchen. "News Focus." *Science*, vol. 304, April 9, 2004, pp. 192–194.)

However, there clearly needs to be a great deal more work in this area to determine which stem cells are best suited for particular types of regenerative therapy. There has also been criticism of these techniques because it has been difficult to demonstrate that any significant fraction of the injected cells has permanently survived. Some research teams are simply using bone marrow cells (presumably containing stem cells), while others are trying to isolate and inject specific cell types. No one has yet tried to use embryonic stem cells, which have the greatest capacity for differentiation. In addition, the use of stem cells in these cases are for a specific disease—post myocardial infarction with significantly impaired myocardial function (heart failure). The use of stem cells for the treatment of aging (sometimes called rejuvenation) is only at the theoretical stage. However, this is an exciting area of research with considerable potential, and only more research will reveal the value of this type of approach.

In fact, embryonic stem cells probably have the greatest potential of solving the **Methuselah Problem**. If embryonic stem cells could be developed in large numbers and given intravenously, and if they localized all over the body, and differentiated into the cell of each organ system, they could, in essence, rejuvenate the body. If this scenario were possible, the possibility of living a thousand years would be right around the corner. In essence, we could simply replace old cells with young new cells. However, preliminary data indicates that it is not going to be that easy. One of the experiments to determine that an embryonic stem cell line has been established is to inject the cells into a "nude mouse." A nude mouse is one that has been bred to have a defective immune system that will not attack and kill foreign cells. They also do not have any hair, hence the name "nude mouse." When "nude mice" are injected with embryonic stem cell lines, they do differentiate into many different cell types, such as bone, nerve cells, muscle cells, etc. This obviously proves the potential for embryonic stem cells to differentiate into all types of cells, and their potential to literally rejuvenate the recipient.

However, it is not clear what "signals" the cells to develop into which type of cell. In addition, the cells do not move to all of the organ systems and tissues of the body and differentiate at that local spot. The cells just stay in a mass of differentiated cells with bone next to nerve cells next to muscle cells, etc. However, a research team at the University of California, San Francisco has identified 22 proteins essential in maintaining embryonic stem cells in an undifferentiated state, and it is not clear that cells injected into nude mice were not already starting to differentiate. If it will be necessary to differentiate cells in vitro, and then inject them at a particular location, regeneration is going to be a long way off. It would work for certain diseases like congestive heart failure, where myocardial cells are injected into the heart (as discussed above), but it will be unlikely that we could do that into every muscle or organ system in the body.

For example, researchers at Baylor College of Medicine (my old school, where I was on the faculty and taught from 1980 to 1985), have found a gene (Sox17) that is necessary to channel primitive mesodermal cells to cardiac mesoderm. Stem cell therapies will probably provide the answer for transplantation therapies and will be a whole new approach to treating many diseases. However, whether regeneration and reversal of aging can be accomplished with stem cells will need to wait for future development. Working with embryonic stem cells is in a very early stage of development, and I am sure that we have a great deal more to learn.

Aging at the Level of the Organism

So far, we have really concentrated on aging at the molecular and cellular level. In fact, aging can also be described at the tissue or organ level, and by a group of "aging diseases" such as Alzheimer's, Parkinson's disease, and others. Even atherosclerosis and cancer (at least in some respects) can be viewed as "aging diseases." However, most of the effects of aging of organ systems and the body are cellular (and even molecular) effects viewed at the level of the whole organ, such as the heart or brain. With increased organization—cells organized into tissues, tissues organized into organs, and organs organized into animals—there appears to be a "domino" effect caused by aging. As one cell type fails, the tissue and organ system composed of those cell types begins to fail, and this affects other organ systems which then function poorly and adversely affects other organ systems including the originally failing organ, etc. The net effect is the animal ages.

Many organ system changes occur with aging, including a decrease in muscle mass and function, decrease immune system function, decrease in hormone production and the list goes on and on, and is quite extensive. In fact, aging literally affects every system in the animal. All these changes are clearly a part of the aging process, and there is considerable literature on the effects of aging at the level of organ and tissue levels. However, none of these changes have been shown to be what might be considered a "primary" cause of aging, and again most of these aging effects appear to be explainable in terms of many of the more basic processes that we have been describing. At the level of the organism, aging appears to be the effect of the accumulated cellular and molecular changes reflected at the level of tissues and organs, which, again, is a result of changes at the cellular level.

It is also not yet clear whether prevention or particularly reversal of the aging process would reverse any of the "aging diseases" such as Alzheimer's or Parkinson's. However, looking at the basic mechanisms of the diseases, it is highly probable that separate solutions will be necessary. In other words, the problem of aging diseases (once they have occurred), will probably not be reversible simply by correcting the biochemical problems of oxidative damage or the genetic programming problems associated with short telomeres. The damaging effect of these problems will either have to be prevented in the first place or will require additional therapies specifically aimed at the damage done by each disease. For example, the calcified plaques in arteries caused by atherosclerosis (which causes heart attacks and strokes) are unlikely to simply dissolve even if methods to stop oxidative damage and extend telomere length and some other wonderful anti-aging treatment is developed.

Current Aging Therapy

To start with, the short (and correct) answer…there are no **proven** effective treatments that have been shown to reverse the aging process in Humans. (However, significant caloric restriction has been shown to potentially slow aging (see below). In fact, despite many interesting experiments of prolonging the life expectancy in yeast, C. elegans, fruit flies, etc., and many billions of dollars spent on vitamins, herbs, hormones, and other therapies, we are nowhere near solving the problem of aging. The ironic fact is that if those billions of dollars were spent on real aging research rather than wasted on the seemingly limitless and useless "snake oil" preparations claimed to slow aging, a real and effective approach to slowing the aging process might have been found within the lifetime of many who have wasted their money on financing quacks and charlatans. It is even clearer that there is no process currently available to reverse aging—often called rejuvenation. Again, the prospect of "rejuvenation" or age reversal would be a lot closer to reality if people donated their money for legitimate research in this area rather than spending their money on all the current nonsense that is being passed off as rejuvenation remedies. Unfortunately, as we have continued to emphasize, Humanity is not nearly as intelligent (as a group) as many continue to espouse.

However, there is some data suggesting that caloric restriction can significantly extend lifespan. Caloric restriction of 40 to 80% of a normal diet has been suggested to increased lifespan and healthspan by as much as 20 to 80%. This has now been demonstrated in many different animal species, including C. elegans, yeast, and mice, and appears to be a real phenomenon. However, this has not yet been demonstrated in man. Also, this is clearly not an answer to the **Methuselah Problem**. Although it does appear to be a valid approach to adding a few years of longer life, we clearly need to look for a better answer. In addition, such an approach of chronic semi-starvation is not particularly appealing to most individuals.

The underlying biologic basis for increased lifespan secondary to caloric restriction is not entirely clear. However, as noted above, there are reports that link caloric restriction to the Sir2 family of genes. It has been suggested that caloric restriction "slows" metabolism and reduces oxidative damage. However, the rate of metabolism (also known as the "rate-of-living" theory) has many glaring exceptions. For example, birds have metabolic rates twice as fast as mammals, yet some birds live much longer. Parrots have been known to outlive elephants. Hummingbirds can live for as long as 14 years, which based on metabolic rate (interpreted here as energy consumption per pound) is equivalent to a Human living to 500 years of age. (Wright, Karen. "Staying Alive." *Discover*,

November 2003.) In fact, Humans already live four times longer than we should based on our size and metabolic rate compared to other mammals.

Leonard Guarente, who we previously discussed in terms of the SIR1, has suggested that the SIR1 gene may play a critical role in the lifespan extension noted in association with caloric restriction. In caloric restriction experiments on yeast, Guarente and colleagues noted lifespan extension only if SIR1 (and cofactor NAD) were present. If SIRT1 was deleted (or its cofactor depleted), no increase in lifespan was noted with caloric restriction in mice. This suggests that caloric restriction may be functioning through SIR2 (at least in mice). Similarly, SIR2 may also play a role in increasing lifespan in C. elegans, but by a different mechanism. (www.hms.Harvard.edu/armenise/old_site/4b_link.htm) There are also other members of the SIR2 family that appear to be involved with life extension through caloric restriction. Strategies based on perturbing SIR2 by drugs or genetic modification might be an approach to gaining the aging advantages of caloric restriction without starvation.

Although studies involving caloric restriction may point to some underlying mechanism in the basic biology of aging (perhaps the SIR2 gene or its homologs in mammals), caloric restriction per se or slowing metabolic rate (short of putting someone in a deep freezer) is probably not going to be the answer to the **Methuselah Problem,** and it is very doubtful that many people are going to be willing to live on a semi-starvation diet for most of their lives to achieve a few years expansion of their lifespan. Therefore, barring some additional data or basic breakthrough in understanding based on caloric restriction or slowing of metabolism, these are probably not the pathways to a solution of the **Methuselah Problem**.

Since the Human genome has been sequenced, there has been some reasonable expectation that the genetic pathways involved in functions of growth, metabolism, and cellular maintenance, repair and replacement will soon be worked out. For example, in April of 1996, the genome sequence for yeast was completed, and it is anticipated that the primary function of most of the approximate 5,800 proteins produced by yeast will soon be worked out. (http://www.aeiveos.com/issues.html) Similarly, the gene sequence of the nematode worm, C. elegans has been published, and scientists are hard at work determining the function of the proteins produced. This has raised the hope that the genes and proteins involved in various aspects of aging will also be rapidly determined. The estimates of the number of genes involved varies from dozens to hundreds. Although the optimistic anticipation is that manipulation of these genes will result in the ultimate solution to the **Methuselah Problem**, this may be a more difficult genetic biochemical problem than most anticipated. We should recall that the estimated number of gene differences between a mouse and man, which separated about 80 million years ago, has been estimated to be only about 300. (http://www.genome.gov/11511308) In other words, manipulation of roughly 300 genes could, in principle, change a mouse into a man (or vice versa). Yet, most scientists would, I am sure, consider such manipulation a formidable task. The manipulation of the dozens to hundreds of genes involved in aging may also be quite formidable. Although the specific genes and their involvement in aging may be delineated, the modification of that number of genes may be difficult (at best) when dealing with the fertilized zygote, and perhaps impossible at the level of trying to genetically engineer hundreds of genes in an adult. In fact, it is beginning to appear that aging is going to involve many genes and many mechanisms, which is probably the worst-case scenario in terms of solving the **Methuselah Problem**. Aging is not one problem, but many problems, each with a different solution. We are now at the very early stage of simply figuring out: "what are the problems?"

The last "solution" to the aging problem that we will briefly consider is cryogenics. This is the current very expensive process being marketed in which people and/or their brains are frozen shortly after death. They are then maintained in a frozen state for decades or supposedly even centuries. The concept is to thaw them only after science has advanced to an appropriate stage where the frozen individual's medical problems and the problems associated with freezing have been solved. Sounds…reasonable, and certainly a few rich and famous people have opted to try this approach to immortality. Cryogenics is a real science, and there has been a considerable amount of credible work done in this area. Bacteria and even Human cells have been frozen for years, and later thawed with good recovery. Early Human (and other animals such as cattle, etc.) embryos have been frozen for years and later thawed and implanted with the birth of healthy newborns. Therefore, the science of cryogenics is real, and continues to advance. In addition, one of the basic premises behind this approach to aging—that science will continue to advance to the point that many of the current medical and biological problems will eventually be solved—is almost a given or at least will be a very good bet in the future. However, how long in the future is at present a big guess.

Before you rush out to have your head frozen, there are a lot of problems with this approach. Kevin Miller has referred to freezing as: "Immortality as a Popsicle." The first and critical difficulty lies in the "thawing problem." There is no current method of successfully thawing people or brains) that have been frozen. In fact, the solution to the "thawing problem" may (and probably does) lie in the freezing technique. In other words, when and if the "thawing problem" is ever solved, the solution may lie in the technique used for the initial freezing. If you did not have the necessary freezing technique, then thawing may be impossible. If you were frozen using an "early freezing technique"—like today— it may never be possible to successfully thaw that individual. We have begun to apply the technique of freezing people and brains before we know how to thaw them.

But there is still another ever more practical concern. If it takes three centuries to solve the necessary medical, biological and cryogenic problems to allow some of the current frozen individuals to be thawed, will these current companies who are collecting money to preserve frozen individuals still be in existence…faithfully maintaining someone's frozen head. Don't count on it. Once the money is gone, the company will be gone, and that is a more fundamental principle of business than anything in the basic science of cryogenics. It is called economics—count on it. Economics may be the death of the current cryopreserved faithful.

Another problem with current cryogenic approaches is that it is becoming radicalized into a religion. What is needed is science, and what we are getting is religion laced with rampant profiteering. Kevin Miller quoting cryobiologist Dr. Kenneth Storey has expressed the current situation nicely:

"Cryonics is 'more or less a theology,' Storey says, 'there is really no difference between cryonics and any other religious organization. They have the truth with no proof; you must have faith, but you can never see a real example of it; you must do what they say without any hesitation (give large amounts of money to them every so often); and they have the key to eternal life.'" (Miller, Kevin. "Cryonics Redux" *Skeptic*. vol. 11, No. 1, 2004.)

It may be true that your cells will not degenerate while frozen but spending an eternity as a popsicle (and never being able to be thawed) is not a solution to the **Methuselah Problem**. This is not to state that freezing and successfully thawing a mammal will never be done. It is simply to say that it cannot be done **now**.

Can we realistically expect the **Methuselah Problem** to be solved, and… when?

The prolongation of Human life expectancy or perhaps even the reversal of the aging process may not be as simple a scientific problem as we have tried to demonstrate. There are no examples of large animals that live a thousand years. If we are considering extending the lifespan of mankind to a thousand years, we are talking about changing the basic biology. Of the four major areas of future prophecies that we are discussing, the **Methuselah Problem** is probably going to be the most difficult to solve. However, there is nothing in principle that should preclude this prognostication from becoming a reality. The problem appears solvable; however. it is worrisome that no multicellular creature (with the possible exception of some trees) have solved the problem in the past 600 million years. We might ask why? If most creatures lived 200 years, then evolution would have been slowed down considerably. Evolution thrives on passing down mutational changes to offspring. However, that answer has built into the improbable requirement that there is some guiding hand (a God) that is directing evolution toward some goal.

In fact, because of the factors we previously discussed regarding Moore's law and the Malthus Doctrine—men actively working to make something happen (Moore's Law) or to prevent something from happening (Malthus Doctrine), and the considerable advances in basic biological science, a solution to the **Methuselah Problem** will hopefully ultimately be found... realistically. We should expect significant progress in this area within the next 100 years. You and I will not see it. We have come a little too early. It is possible that some real progress could be seen within our grandchildren's lifespan. Although, the lifespan of man almost doubled in the 20th century that was due to the conquest of many specific diseases—especially infectious diseases, and not just by medicine The extension of the Human lifespan (especially the healthspan) beyond 120 years will require more than improved medicine. It will require a basic understanding of the aging process and changes in basic biology. In fact, as we have previously noted, medicine is unlikely to make major advancements in extending Human lifespan in the future unless there is a change in the basic paradigm with an emphasis on aging research.

Considering the technical capability and knowledge base of science and medicine at the beginning of the 20th century (they could be considered primitive...compared to the beginning of the 21st century), we might anticipate considerable scientific progress in the 21 century. I am reminded of Kurzweil's prediction of a doubling of biotechnology knowledge and techniques every ten years in the 21 century, which translates into 140 years of progress (by 20th century standards) by the year 2030. We will probably need a lot of scientific breakthroughs to finally solve the **Methuselah Problem**... many of which are not yet even on the drawing board. Unquestionably, seeking a real scientific answer to the **Methuselah Problem** would have been an unrealistic scientific dream at the beginning of the 20th century. We simply did not have the basic science background in biology or the technical capability to realistically attack the problem at that time. Now...at the beginning of the 21st century, the **Methuselah Problem** can be viewed as a realistic scientific area of research. Scientists have recently just begun to seriously work on the understanding of aging, and the possibility of seeking solutions—to slow or even reverse the aging process. The **Methuselah Problem** could be compared to the "problem" of flying as viewed by men of the 18th century. It was a dream to think that men could ever fly...in the 18th century but flying became a reality in the 20th century when engineering and technology became adequate to the task and the goal was actively pursued and...achieved. In fact, the "flying problem" has been solved well beyond the wildest dreams or possible predictions of 18th century man.

The answer(s) to the **Methuselah Problem** probably (hopefully) can be solved by the end of this millennium—perhaps a life expectancy of **1000 years by 3000 B.C.E.**, and significant progress may be a reality within 50 to 100 years. The exact timing is impossible to predict. We cannot, at this point, simply apply Moore's law to aging as was done with computing. In the "Moore's Law" prediction, the problem was well defined. In the aging process, the problem is not yet even fully defined. The problem is not just aging. Aging is the consequence of some (perhaps many) other biochemical, genetic and cellular processes, which must clearly be defined.

Once the **Methuselah Problem** is solved, the possibility of Human immortality would then be right around the corner. I use the term relative immortality to signify that there would be no programmed biologic death—there would be immortality in principle. Of course, if someone drops an atomic bomb on you—forget about telomeres, genetics, or immortality in principle. Albert Einstein once said (year 1946): "There is no foreseeable defense against atomic bombs..." (http://paulingexhibit.org/exhibit/atomic-era.html) Mortality by mayhem will always be with us, and it will, in the far future, become the limiting factor on

Human life expectancy. Current man and his predilections for wars and murders would clearly be a limiting factor to any biological immortality even in principle.

We are too close to the Homo sapiens that fought their way out of the jungles. This is one of the reasons that we have been emphasizing the fact that current "modern" man is a **work in progress.** Unfortunately, as we have mentioned, we are also the "sacrificial generations…the Moses generation." We see the goals…but they are just out of our reach. **Intelligent Directed Evolution (IDE$_H$)** for man will be needed to bring these goals within our grasp. In at least some respects, the solution to the **Methuselah Problem** is intimately tied to solutions to the other problems we have been discussing such as genetic engineering—the remaking of man, and what we will shortly be discussing—increased intelligence.

Gregory Stock has noted three possible approaches to the aging problem: 1) slowing of aging in adults, 2) reversal of aging in adults, and 3) germline changes. (Stock, Gregory. http://research.mednet.ucla.edu/pmts/sens/transbody.htm) Slowing of aging in adults is basically what is currently being tried—antioxidant vitamins, caloric restriction, hormone therapy or the acetyl-L-carnitine/lipoic acid, and probably many more to come in the future. These are the current common approaches that we have mentioned. However, these are basically "Band-Aid" techniques that may (or may not) add a small amount of time to the Human lifespan. They are not approaches that attack and solve the "core problem" in aging. In fact, at the present time, we really do not fully understand what is the "core problem" in aging. Yes…telomeres/telomerase, oxidative damage and mutations, hormones, mitochondrial function, and longevity genes all appear to contribute to aging, but are these really the "core problems?" If we completely corrected the "mitochondrial problem," would the average lifespan be extended to a thousand years—almost certainly not.

As noted above, the "worst case" scenario would be that there is no "core problem"—aging is a bunch of "little" problems, and each will have to be addressed separately. We age because the whole biological system begins to collapse with each problem creating additional problems—a domino effect like what we have discussed, but at the molecular level.

Stock's second approach to the aging problem involves reversal of aging in adults (regeneration). Although there has been a great deal of talk about this prospect, there is no evidence that this has ever been done in Humans, or in any other animal. Probably the current "best bet" in this area would be stem cell research.

The last approach is genetic engineering of improved Humans—getting rid of the bad genes and adding new genes, which we have noted will begin within the lifetime of most people under 30 years of age. These techniques will be needed to make germline changes that may be needed for any significant progress in the area of aging. However, like the genetic engineering of Humans in general, a solution to the **Methuselah Problem** will probably require many small steps. We may literally have to slug our way through this problem.

However, we are at a level where we can seriously and realistically begin to pursue…the impossible immortality. Is it blasphemy to pursue immortality…not really, man has been doing it since the dawn of civilization (and perhaps before as reflected in ritual burials and the Egyptian afterlife). Life after death (Heaven) is nothing more than one approach to immortality. We are now only changing our technique of attaining immortality. Instead of reading prayers and spells from the Egyptian Book of the Dead to guide us in the underworld of life after death, as envisioned by early Egyptians, we will be using scientific knowledge and research. We will reach immortality not by praying, certainly not by "going to Heaven," but by scientific inquiry. But we must not fear uttering the word "immortality" or to pursue the goal through research. **If we want this goal…no one is going to give this to mankind…but mankind himself.** Failure…only our graves await at the end of the road of failure, but they were there already.

Probably the most important advancement that we have made in aging is not the telomeres or the genes so far discovered associated with aging or even oxidative damage, but the **realization** that aging is a scientific problem that can realistically be pursued…and solved with the technologies at hand or within our grasp.

However, when and if immortality is ever achieved, the sign on the accomplishment will read: "Made by Man."

Chapter 5

THE WISE MONKEY

But are we wise enough?

Intelligence is the "Essence of Man"

As we have learned on our trip, we are Homo sapiens sapiens (HSS). "Sapiens" comes from Latin and means to "be wise." We have literally named ourselves not just the wise, but the wisest of the wise monkeys. Is our name another example of Human conceit—possibly the ultimate self-complement? The name is perhaps a bit immodest, but not entirely without at least some justification. This perspective is not being immodest or arrogant; it is simply a statement of fact—as we see intelligence. When something is true, it is neither arrogance nor conceit; it is simply a statement of fact—perhaps not too humble, but a statement of reality, nonetheless. After all, we are the most intelligent creatures that we know on Earth or in the whole Universe—at least presently. There is clearly no other creature on Earth that even comes close to our intelligence. In fact, intelligence is the defining characteristic of Homo sapiens. As we have repeatedly noted, "advanced" intelligence is the **essence of man**. It is literally our defining characteristic.

Yes, we walk upright on two legs, and we have hands that are very adept at manipulating objects like tools, and we have language (which is also a component of intelligence), but **the essence of mankind is our intelligence.**

From man's intelligence flows all the other characteristics of "Humanity" that literally makes mankind what we are. Every other characteristic of man is either shared with other creatures in at least equal portions or can be found in other creatures in even greater proportions or in an even more refined character. Whether we are considering size, speed, muscle strength, vision, or more subtle characteristics such as aspects of cellular metabolism, mankind generally does not get the Olympic gold medal. We have only one gold medal in the contest of life—advanced intelligence.

However, recognition of man's special ability does not preclude at least some intelligence in other animals or even in entirely different types of creatures (extraterrestrials) or even non-biological entities (like computers). In addition, recognition of man's intelligence at the present time does not preclude the development of much greater intelligence in the future.

In fact, one of the most important points that we will be emphasizing is that intelligence should be viewed in a broader perspective in order to have a better outlook in where mankind might be "going" with intelligence in the future or perhaps even where intelligence might be going—with or without mankind. However, this is problematic for at least two reasons: 1) We do not entirely understand intelligence, particularly at the level of the neuron and neural circuits in the brain, and 2) it is particularly difficult to project or generalize what a "higher intelligence" would be like, since no "higher intelligence" presently exists. Predicting or characterizing marked advancements in intelligence that we are going to undertake momentarily becomes even more problematic since Humans are the only example, and there is not another example of advanced intelligence.

However, one of our basic maxims is that man is a **work in progress,** which we take to imply that we are now different from what we were in the past, and we are different from what we will be like in the future—including changes in our intelligence. Roughly translated, mankind is in transition, and always has been and probably always will be. In fact, the entire cosmos and everything in it is also in transition. If we find a 3.5-billion-year-old rock that is essentially the same as when it was formed, we might want to question the concept of universal transition (evolution). However, in the case of the rock, we are really talking about the rate of change—not the fact that change is occurring. If we expand our reference time frame (to a geologic time scale in millions or even billions of years), the rock can be seen "in transition" just like everything else. Although there has been considerable "resistance" to the concept of Human evolution or the evolution of life in general, everything in the Universe is moving and everything is evolving. Stars and our Sun have a "life cycle" and they are evolving. The quantity of each element in the Universe (carbon, iron, etc.) is changing. Everything is moving.

We are treating this premise of continual change (man as a **work in progress**) as a tautology—especially when it is applied to a biological entity, and particularly when applied to mankind. We looked at some of the factors that might stop changes in man in the chapter *Armageddon...*, and almost all involve extinction. Extinction is about the only exception to continued biologic change. As long as man exists, we will be in transition—a **work in progress**. However, we should not consider this a bad omen from God! Evolving is how we got to where we are, and it is a good thing! In fact, the real essence of mankind may lie in the future! Do not fear the future because it sure as hell cannot be any worse than the past. This is particularly true if mankind and intelligence is finally a part of the future of evolution.

Not surprisingly, as the **essence of man,** intelligence should be one of the most important areas of future Human development (change or transition)—just as occurred some 150,000 to 200,000 years ago when Homo sapiens evolved from Homo erectus (note, there is still considerable argument amongst paleontologists as to exactly which species was mankind's last common ancestor or even the time period of exactly when that occurred. We will simply avoid the controversy and the multiple candidate species (such as Homo heidelbergensis), which is continually changing, and use Homo erectus as our "stand in" for our last common non-sapiens ancestor until this controversy is better resolved). During that period of transition, the brain of Homo erectus expanded from a cranial capacity of about 800 to 1000 cc. to about 1,350 cc. for modern man. It is not clear whether the Homo erectus to Homo sapiens transition was a slowly evolving transition over hundreds of thousands of years in a "classical" mutation-selection manor of Darwinian evolution or a much more rapid process (we will discuss more aspects of shortly). As we have previously noted, genetic analysis and comparisons of genetic variations between Humans and comparisons with other species do indicate that man developed from a relatively small group of ancestors. Although it is doubtful that a baby was born carrying a single gene mutation that produced Humans, considering some of the experiments we will look at (such as rapid transition), it is at least possible. Again, we will be looking at such experiments that demonstrate rapid change in brain size from a change in one gene and rapid change in some aspects of intelligence such a processing speed shortly. However, regardless of how many mutations or genetic changes were required or over how long a period they have occurred, the most important changes on the road to Human development have involved changes in intelligence. These changes have literally defined mankind.

Nevertheless, with the development of **Intelligent Directed Evolution** that we have discussed (using Human intelligence combined with genetic engineering technology [**IDE**$_H$]), the changes in evolutionary dynamics now have the potential to be incredibly rapid. It may no longer take hundreds of thousands to millions of years to "evolve" significant changes in a species. Significant changes could be produced in a few decades, and a century may be all that is necessary to totally remake a species—even mankind. The third millennium could well be man's time of epiphany with our future selves. The remaking of man will probably not involve radical physical changes. As we have previously noted, we are relatively happy with our current morphology. Also, as we have noted (from the Bible), we are made in the image of God…so why change! Since man will literally be directing these future changes, it is relatively easy to make some "educated guesses" as to what those changes will be. As we noted in our discussion of genetic engineering, the future changes or evolution of Humankind will probably center on the elimination or modification of genes and alleles that are associated with diseases, decreasing or elimination of aging, and the probable manipulation of designer genes that will make us all look like a Bo Derek "Ten" (or whatever phenotype is in fashion at the time). However, as man remakes man, one of the most important target areas for future fundamental development—evolution, will undoubtedly be an increase in intelligence. IDE$_H$ could cure all genetic diseases, give man incredible resistance to all known current diseases, eliminate atherosclerosis (responsible for heart attacks and strokes), eliminate cancer, Alzheimer's, etc., provide incredible strength and stamina, and extend the lifespan of man to that of the biblical Methuselah; however, if the intelligence of man is not significantly enhanced, these other "accomplishments" or "advancements" would be almost meaningless in terms of any significant evolution of Homo sapiens. We would simply be stronger, healthie, and longer-lived Homo sapiens, but we would have changed (advanced or evolved, if you will) not at all. Without a significant increase in intelligence, we would have "evolved" ... nowhere! It would be as if Homo erectus wandered the Earth for two million years, and evolved to look like Homo sapiens, but retained the same Homo erectus brain with the same intelligence.

In addition, with man's increasing technical skills, if there is not a significant increase in intelligence, including social intelligence, "personal intelligence," Theory of Mind, and basic intelligence functions that govern Human interaction, there is a considerable probability that man will eventually destroy himself and/or his habitat. Although many problems such as wars, crime, social injustice, and the ecologic havoc that we are perpetrating on our planet are generally not perceived in terms of a function of the intelligence of the Human population of the world, I would opine that these problems (and many others such as the magical thinking and mystical beliefs that are used as the guiding principles of most of the seven-plus billion Humans currently living on the planet Earth are in fact **a direct result of man's limited intelligence.** We must begin to appreciate that there are intelligence functions (or perhaps better characterized as a lack of sufficient intelligence functions) involved in our social interactions and belief systems and even in our emotional responses. Viewed in this manner, advanced intelligence is a vital factor necessary for future Human survival in a complex Universe, as we will now consider. In other words, developing much more advanced intelligence is not a luxury; it is an absolute essential prerequisite for the future survival of mankind.

Unfortunately, mankind clearly does not recognize his current limitations. Rene Descartes (1596–1650), the great French mathematician, philosopher, scientist and thinker who gave us the logic proof of existence: **"I think, therefore I am,"** also made the interesting observation that: **"Of all things, good sense is the most fairly distributed: everyone thinks he is so well supplied with it that even those who are the hardest to satisfy in every other respect never desire more of it than they already have."**
(http://teren.wikiquote.org/wiki/Ren%C3%A9_Descarte)

We all think that we are already smart enough. This is the Achilles heel of mankind, and as we have opined, the quickest road to Human extinction. There are even powerful institutions that are absolutely opposed to any changes in mankind's intelligence!

Intelligence and Survival

Several pundits have suggested that life has an evolutionary imperative for intelligence. In other words, the development of intelligence was required (or planned—usually by a God or some cosmic power) as an integral part of the evolutionary development of life. However, this concept confuses the incredible pragmatic **utility** of intelligence with an implied planned decree or edict to create intelligence. The increasing intelligence noted during evolutionary development occurs almost in parallel with phylogenetic development, but it is not the result of any grand design on the part of "Mother Nature" (or God, if you prefer). There is no data other than wishful thinking to support either the Strong or Weak Anthropic Principle or that the world was created so that intelligence would develop (this is a variation of the anthropic principle that has been called the Final Anthropic Principle, which we have previously discussed). However, the rise of intelligence that can be seen throughout biology (phylum through phylum [in the animal kingdom]) is also no accident, neither does it require divine intervention for an explanation. It is simply that the neuron—a cell to control other cells (like muscles) and to make logic choices and allow learning and memory, and to allow an organism to "sense" and then interact with its environment by cell-cell communication between neurons—is one of the greatest inventions of evolutionary biology. Why? Because the function of neurons allows logical interactions (changes), between the organism and the environment. The ability to respond to a changing environment and to learn and remember is at the very heart of survival, and survival is what evolution is all about. Intelligence (in its generalized form to be discussed below) is simply one of the most powerful "tools" in biology for survival. Intelligence in its full breadth and scope is such a powerful survival instrument for life that it's development and enhancement in one form or another was almost assured (this includes even very basic forms of "intelligence" to be discussed below). Intelligence may, in a sense, be "built into evolution," not because "nature" has any intrinsic interest in intelligence, but because intelligence is one of the two most powerful and most basic survival factors available to biological systems. On a basic level, the ability to recognize danger and to simply crawl away demonstrates significant intelligence, and the survival advantage is considerable. Although I am very conscious (and sympathetic) of the general concepts of intelligence involving language, logic and mathematics, learning, memory, etc. (mixed in appropriate concentrations, and embedded in a machine to mix, match and express these various components), I am also appreciative of the fact that intelligence begins as rather basic functions that have been critical to the survival of animals. A neuron and the synaptic connection between neurons (to be discussed below) is the basic "unit" of intelligence. Many recognize only advanced forms of behavior or "thinking" as intelligence. However, intelligence is clearly a spectrum beginning at zero and progressing higher, and like many continuous linear functions, it becomes difficult to separate two closely approximated points on a line. The difference between high and low intelligence on such a continuum is a matter of perspective. We will return to this point shortly when we discuss the absolute intelligence scale.

It is the survival advantage of intelligence that has led some to the conclusion that intelligence is built into evolution (as a biological imperative). Intelligence has even been suggested as the ultimate "purpose" of evolution as expressed in the Final Anthropic Principle and more subtly implied by the Weak and Strong Anthropic Principles (but to different degrees). There is a clear increased progression of intelligence with biologic development, but this is due to its incredible pragmatic usefulness, and not due to some mystical imperative. The cosmic imperative for intelligence is simply its incredible evolutionary utility—nothing more. Intelligence is such a useful survival "tool" that it appears to be ordained (I smell a God), but it is only ordained by its utility which enhances survival.

(Note: The second important survival technique is reproduction. The two most important biological survival strategies are to "out reproduce" competitors or hostile environmental conditions [bacteria are an example] or "outsmart" predators, prey and hostile environmental conditions [which we will generalize as an intelligence function].)

However, it is also obvious that there are other "techniques" of evolution that have allowed survival without any intelligence and without any nerve cells. The entire plant kingdom has survived quite well without nerve cells and without intelligence. Plants clearly show that there is no cosmic imperative requiring intelligence for life or for survivability or even for evolution. In fact, the example of the plant kingdom appears to place an important limitation on the necessity for intelligence in lifeforms. The limitation appears to stem from a "free lunch."

Plants obtain their energy from sunlight through photosynthesis in plastids; they (plants) get a "free lunch." However, most of us have been taught that there are no real "free lunches" in life. In the case of plants, their "free lunch" **cost** them the lack of development of nerve cells and absence of intelligence (also a lack of purposeful mobility, which is based on muscles, which is controlled by nerve cells). The development of intelligence has been limited to "predators," which must go out and work for their lunch (I am using the term predator here very broadly to include any metazoan that "eats" other organisms, including herbivores). In other words, the reason intelligence developed in "predators" is that we get no free lunch, and we have literally been forced to evolve nerves and brains to go get our lunch. In essence, plants' "free lunch" cost them dearly (in terms of lack of any evolutionary "pressure" to develop intelligence or mobility).

GCOS: "This sounds like the 'short course' in Darwinian evolution."
"So be it!"

As we have noted earlier, sponges, which are the first and most basic metazoan animal phylum, do not even have any nerve cells (and by the criteria we are using here...no intelligence). Yet, sponges are also a very successful animal lifeform, and they have survived (in one form or another) for over 600 million years. If intelligence is such an important "survival tool," how do we explain the success of the entire porifera phylum with 5,000 extant species, and their survivability for over 600 million years? Again, the answer appears to lie in a variant of the "free lunch." The "free lunch" in the case of sponges lies in their simplicity. They are predators only in a very limited sense. However, the "higher" we go in predatory phyla, the more important intelligence (and nerve cells) becomes as a "survival tool."

Since intelligence is such a powerful survival "tool," we can be assured that man as a **work in progress** and the current bearer of life's banner of higher intelligence will evolve in the future in such a way as to increase intelligence, and therefore survivability. This is almost a "given," if future evolution is mainly driven by intelligence directed genetic engineering—**IDE**$_{ri}$ (as we have previously conjectured).

Some may object to linking more advanced intelligence to survivability or to consider intelligence in the broader framework being utilized here. In fact, Humanity has not survived because of our ability to do calculus, play chess, recite poetry or use any "higher" intelligence functions.

In our previous discussions of possible Armageddon scenarios, we noted a considerable number of dangers that face Humans (and life in general) in the form of asteroid or comet impacts on Earth, but there are many many more. We have only learned relatively recently that a meteoroid impact was responsible for the killing (extinction) of the dinosaurs about 65 million years ago, and that such impacts are common—relatively speaking on geological time scales. There have been multiple massive asteroid/comets impacts on the Earth in the past, and multiple mass extinctions in the past (from impacts and probably other factors as well, such as severe changes in environmental conditions (weather), as we previously discussed in the chapter *Armageddon...*

However, the dinosaurs were never aware of the danger of asteroid or comet impacts. The dinosaurs had small brains and limited intelligence. They recognized only "local" threats such as a predator, lack of water, food, etc. In fact, the most intelligent creature on Earth (that be us) has only very recently become aware of the danger of asteroids or other impactors. It can realistically be said that it took man over 150,000 to 200,000 years to even begin to understand some of the great dangers that the species Homo sapiens faces. Like the dinosaurs, man has only recognized the "local" dangers very recently. We are only now beginning to recognize some of the other real dangers to Human existence. Asteroids and other impactors are clearly not the only danger facing mankind, and there are probably many other serious dangers that we are completely unaware of at this time with our current level of knowledge (and intelligence).

The answers to some of the currently recognized dangers are not solvable with current science and engineering capabilities... and I would add current intelligence. The survival answers to huge impactors, planetary nebula, or massive weather changes such as another "Snowball Earth" will require incredible scientific, engineering, and organizational advancements...and the cosmic imperative behind solving these and myriads of other problems will be the same incredibly evolutionary survival tool that has gotten us to this point in our journey—intelligence. However, it is going to require a lot more intelligence than is currently allotted to Homo sapiens, because the problems for the long-term survival of man or the Earth are much more complicated than a local predator. The solutions to these problems in the future are going to require advanced mathematics, science, and advanced intelligent thinking and (very important) much more advanced social skills (which we will find later is really a component of intelligence).

In fact, it is ironic that the greatest and most obvious threat to mankind is aging and death (as we have discussed in the preceding chapter), and it is only now beginning to be recognized as the extreme threat that it actually is for every Human being. Why has it taken mankind so long to begin to seek solutions to a threat that has killed every Human that has ever lived? We have recognized the "threat" of dying since recorded history, and probably much earlier—we are all going to die. However, we are only now beginning to recognize that death and aging may not be inevitable—at least in the short time frame of the current Human lifespans. In the past, our solution to the "inevitably of death" has been some variant of denial. The afterlife of the Egyptians, reincarnation, Heaven, the happy hunting ground or some other mythological solution that, in essence, denies that we will cease to exist. We will age and die (it would be hard to deny that fact, which is constantly confirmed), but we will ultimately survive in some other form and place (there is the Santa Claus myth). These "solutions" are fairy tales. Instead of recognizing reality and seeking the basic knowledge and understanding of the fundamental biology, man has accepted aging and death as an inevitability and evoked mythology (God and religion) as a solution. These types of solutions to such a fundamental "problem" (and, in fact, the whole inability to even clearly define the problems that are involved, and to recognize aging and death as the ultimate "local threat" that they truly represent) reflect a dinosaur-like mentality. Mankind has not been smart enough to recognize aging and death as a threat that may be "solved" with real answers rather than accepting these "threats" as an inevitability of life to which we apply some version of denial. God, Heaven…these are the "shortcut" and simplistic answers to the local threat of aging and death. Had we built thousands of laboratories rather than thousands of churches, we might have an answer to the problem of aging and death…now. Unfortunately, mankind has taken the shortcut on our journey, but instead or arriving in Heaven, the shortcut has led us to Hell. We are all paying for our sins of ignorance, which is due to a lack of intelligence. Even now, the majority of the world's population is still not smart enough to "figure it out." They will all die…literally from ignorance. This is, of course, a part of the **Dilemma of Man** that we have repeatedly discussed.

An important facet of a more advanced intelligence is the ability to reason through data and draw logically correct conclusions and delineate the path towards real solutions to problems. Man has simply not recognized the biology of aging or death as a threat that can be managed and solved in any realistic manor. Yet, this "danger" is our biggest threat as individuals, and it has literally been staring us in the face, since the beginning of mankind. As a consequence of our ignorance, we are still (at this time) a long way from any real solutions to aging and death. Incredibly, a large portion of the world's population still does not recognize either the problem and definitely does not have a clue for an approach to a possible solution but continues to be deluded by mythology and magical thinking. They are still on the "shortcut" journey to immortality. The majority of the world's population continues to "solve" some of the most basic problems of man by praying and performing useless rituals. This is truly a pathetic commentary on current Human intelligence!

Remember from the preface: We remain the cows in the pasture awaiting slaughter. "They feed. They reproduce. Life goes on," but one day...a truck arrives to take them to the slaughterhouse. They are slaughtered, but they never "figured it out." Low intelligence has some very nasty consequences.

Where is the road to salvation...we again invoke one of our basic themes: **No one can save man, but man himself** (that is simply another way of saying **"Made by Man"**). This includes saving mankind from ignorance by advancing man's current limited intelligence. The cost of ignorance is high—you will die because of the ignorance of our forefathers...ashes to ashes and dust to dust. If the mass of Humanity does not recognize mankind's current "local threats of aging and death," it is a direct consequence of ignorance. Mankind will simply continue to be taken to the slaughterhouse. We will all die...literally from our ignorance of ourselves and our forefathers!

GCOS: "The Earth is our Heaven, and the Earth is our Hell! Oh, I say unto you mankind...make of it as ye may or you will continue to wallow in ignorance in your Hell on Earth."

Mankind will either increase our obvious limited intelligence; or we will continue to wallow in Hell. In addition to simply surviving as a species or as an individual, an increase in intelligence and the dissemination of intelligence into the cosmos is one of the major moral goals of man (based on **Human-Centered Ethics** as previously discussed) as expressed within the **Manifest Destiny of Man**. If there is no advanced intelligence "out there," then mankind's greatest mission (and meaning for existence) may be to spread intelligence throughout the Universe. It might be viewed as man's highest moral obligation to extend intelligence into the rest of the Universe—in addition, this should also be viewed as a fundamental survival function. However, it is very questionable whether the advanced intelligence that should ultimately be unleashed into the Universe is the current standard Earthly issue found in mankind. Since intelligence is now literally under man's control through IDE_H (although, perhaps not recognized yet by the majority of the world's population), mankind has a moral obligation (or how about a cosmic command from Gods) to propagate a higher form of intelligence into the Universe. Again, this is one of the major conjectures of the **Manifest Destiny of Man.** If we want meaning and purpose for Human life...we must impose such meaning and purpose (as we have previously opined) on ourselves now and in the future.

It is not my purpose here to belittle mankind's accomplishments. Our accomplishments (with our current level of intelligence) have been impressive. Homo sapiens are entitled to a certain amount of pride. However, we still must understand that our current level of intelligence is very limited. Yes...the greatest intelligence that we are aware of in the Universe, but still...very limited. Intelligence has a long way to go here on Earth.

Frankly, advancing the level of man's intelligence even on Earth is not going to be an easy task, and extending advanced intelligence even into the Milky Way Galaxy (let alone the rest of the Universe) will keep mankind (and mankind's descendants) occupied for a very, very long time. However, this is a part of the **Manifest Destiny of Mankind.** We need to roll up our sleeves and go to work. The task is daunting, but no one is going to do it for us.

Future Intelligence

Although it can be legitimately argued that no Human has yet had their intelligence enhanced by IDE_H techniques, such techniques are becoming available, and have already been used in animals to produce "smarter" mice (to be discussed shortly). Such techniques should clearly not be used on Humans until a great deal more is learned about the function of the brain and the effects of any genetic modification. However, natural evolution has probably already made some genetic modifications in very intelligent individuals (let's arbitrarily define this as an IQ of 180 or above) that have not yet been identified, similar to changes in Human capabilities such as the very muscular German child recently born with a mutation of the Myostatin gene as previously discussed. Probably the first and best initial approach to increasing Human intelligence will be to identify and propagate those mutations or genetic modifications already present in at least a few individuals. Even using such simple techniques that Darwin recognized in pigeon breeding, it should be possible to increase the average intelligence of Humans from an average IQ of 100 to at least 180 or 190, which is the current known upper limits of Human intelligence; however, even that small increase would still be almost a doubling of mankind's current average intelligence.

"Would that be an adequate increase in intelligence to get us where we need to go?"

"Absolutely not!"

However, at least it would be a start on the road to higher intelligence.

What we now "possess" in intelligence is not what should be extended to the far reaches of the Universe and is definitely not adequate for mankind's long-term survival. However, we do appear to have enough cleverness to possibly extend intelligence in the future to the descendants of mankind using **IDE**$_n$. Indeed, one of the mechanisms through which man could eventually fulfill our **Manifest Destiny** would be to create the next more intelligent species of man (we will be discussing some of the other methods in the next chapter, *The Dumb Machine*). In this role, we would be the creator species, but not the actual future higher intelligence per se. However, this is still very consistent with the maxim of mankind as a **work in progress**. We have created many Gods in myth and fable, but the time is rapidly approaching when we will have the technical power to create Gods in reality. In fact, we may eventually fulfill the biblical statement that man was created in the image of God—we will be the Gods or at least the creators of these future Gods. This is something akin to the question in the 1950s: "Are there spacemen?" Yes, there definitely are spacemen; we are the spacemen! In the not-too-distant future, the question: "Is there a God?" will be answered similarly. Yes, there definitely are Gods; **we are the Gods.** Fantasy...not really. Remember when we were looking at the different facets or what might be called "properties" of what has throughout history been ascribed to Gods. We concluded that it was very important to define exactly what it was that we were considering...otherwise we would soon be speaking babble. Gods are generally described as very powerful and all-knowing (read that as intelligent). Those "properties" are a part of the superlative God (all-good, all-knowing, all-powerful, etc.). Gods are the best at whatever is admirable. Then there is also the candy store God that gives us things, and the God of the foxhole who saves us when we are in a bad situation. We are dying of cancer...we call on the God of the foxhole to save us. I would simply reiterate the Clarke/Shermer generalization that a sufficiently advanced intelligence would appear to be all-knowing and all-powerful and could get their own candy and cure their own cancer. In other words, they would be a God. However, I will not try to put an exact intelligence number (**IQ**$_{ab}$ discussed below) on a God, since it is really a matter of definition and arbitrary opinion.

In fact, since there is no real acceptable definition of what constitutes a God, either by intelligence or power or whatever parameter one chooses; if we consider the fact that mankind has developed the only higher or what is sometimes referred to as sentient intelligence, which actually has some understanding of the world in which we live, it is not unreasonable to even consider ourselves as Gods. We should recall that Julius Caesar, Augustus Caesar, and many other Roman emperors (and even some of their relatives) were considered Gods (and had temples built to worship them). However, considering the great deal of foolishness perpetrated by the Human species, if we are Gods (it is a matter of definition), we are clearly at the lowest level possible for any reasonable definition of a God.

However, creating realistic God's is going to require a much more intelligent creature that the current Homo sapiens model. I would opine that man (as we know him) will probably only begin the process. We may be capable of developing the next species of Homo—perhaps to be called by their major characteristic, Homo intelligenticus. Would this new species of Homo be a God? Again, this is all a matter of definition. This child of mankind will have to be much smarter to even begin to be considered as a new species.

"How much smarter?"

Since the intelligence level of our last Homo ancestor, Homo erectus, is unknown, let's instead consider the intelligence of our most closely related species whose relative intelligence level is known. How much smarter is Homo sapiens compared to the chimpanzee? Scientists working with chimpanzees have demonstrated considerable learning ability, memory, and the ability to solve simple problems and use tools. However, our brains are roughly three to four times larger than chimpanzee's. It is difficult to put a number on such a comparison of intelligence, but perhaps men are four to five times "smarter" than chimpanzees (and I am not going to try to define smart other than the generally accepted but imprecise meaning), but arguably, Humans are probably not more than 10 times smarter. Although arguable and arbitrary, I am going to opine that at least a tenfold increase in intelligence would be needed to justify the recognition of a new species distinct from man. A new species in biology is generally defined by its inability to breed with a closely related, but separate species. In other words, a new species has developed enough genetic changes that it can no longer breed successfully with its closely related but different species. However, here we are characterizing a new species by a marked increase in intelligence such as occurred between our nearest living relative and Homo sapiens, and not by the lack of the ability to breed. Such an increase in intelligence may or may not require enough changes in our chromosomes to prevent interbreeding between Homo sapiens and Homo intelligenticus. It will be probable that this new intelligent species will be genetically modified enough that they cannot breed with Homo sapiens. However, there is nothing "in principle" that would require that scenario. With advances in intelligence by IDE_H, it will become necessary to re-evaluate the current concept of what defines a separate species. It is even more controversial to define what intelligence level would be adequate to be considered a God. Our conjecture here is that an intelligence of at least 1,000 (minimum) would be necessary to begin to consider that enough changes have occurred to recognize such a being as a new species. How that might occur or even how such intelligence might be measured is what we will be discussing shortly.

In order to succeed, we (current Humans) must be willing and able to pass the torch of intelligence onto the next generation, who will be different enough from Homo sapiens that they will form a new species—just as Homo sapiens is different enough to constitute a new species that descended from Homo erectus (or whatever the latest new species is that is found and believed to be mankind's last ancestor).

How will we make this new intelligence? Let us first consider several aspects of intelligence that may give us a better perspective as to what may be in store for the future of "mankind," and for the future of intelligence. We need to consider several basic and fundamental questions regarding intelligence.

1. What is intelligence?
2. What is the real basic "hardware" for intelligence?
3. How do we measure intelligence?
4. Can we dissect intelligence into some parts, and distinguish some of the components of intelligence?
5. Is there an "absolute scale" for intelligence, and can we construct such a scale?
6. How intelligent is Homo sapiens on this "absolute scale?"
7. What is the intelligence of other animals on this absolute scale?
8. Are there other intelligent creatures in the Universe?
9. Finally, our real concern here is: "what would an advanced intelligence be like, and can we realistically make changes in the Human brain that would increase our intelligence significantly?"

We have no experience with any higher forms of intelligence than ourselves, and even considering such a question is difficult. How do you describe or characterize something that does not exist? The extent of the problem of comprehending advanced intelligence might be compared to the problem of flatlanders (this comparison may have even more meaning when we get to dimensional thinking below). Flatlanders lived in a two-dimensional world as described by the protagonist A. Square in Edwin A. Abbott's book *Flatland*, published in 1884. The book does several things including satirizing social perspectives of Victorian England, but it also illustrates the difficulty of a two-dimensional "being" interacting or even understanding phenomena in higher dimensions. By appreciating the difficulties of an imaginary creature from a lower dimension in understanding higher geometries like a three-dimensional world, we inherently begin to appreciate how difficult it is for us to understand higher dimensions such as the fourth or even higher spatial dimensions. (Banchoff, T. F., http://www.geom.uiuc.edu/~banchoff/ISR/ISR.html) Currently we generally equate time as a fourth dimension, however, here I am restricting a fourth or higher dimension to space only.

Likewise, it is probably impossible for us to truly understand or appreciate a significantly higher intelligence—it is simply beyond our experience and capability. However, mathematicians have been capable of working in higher dimensions despite the fact that, like flatlanders know, one cannot truly appreciate exactly what higher dimensions would be like. For example, string theory, which has been suggested as the potential "theory of everything" (LOL [Laughing out loud—a common net phrase] …I love those physicists, they are so humble) only works when there are ten or eleven spatial dimensions. Our effort here will simply be a lower dimensional attempt to project phenomena into a higher dimension of intelligence.

We will try to answer the questions of increased intelligence within some of the areas that we will discuss, and also (at the end) try to consider possible types of intelligence not even possessed by man at this time, but of course, our effort will probably be similar to that of the character A. Square in *Flatland*—a two dimensional creature trying to imagine something in (3-dimensions) that is unimaginable within his domain. Let's look at the impossible!

What is Intelligence?

Intelligence is an enigma. We can define it. In fact, the definitions are almost endless. As Humans, we all feel that we have at least some of it—so we each feel that we have some personal experience with intelligence, but no one really knows exactly what it is that we "have." We can (supposedly) test for intelligence, and measure it, which is incredible…considering that we do not really know what the entity is that we are testing for. We know that it is located in our brain, but we do not know exactly where—perhaps nowhere exactly—perhaps everywhere. There have been early attempts to "map" certain functions to specific areas of the brain, but Karl Lashley, an American neurologist, has convinced most neuroscientists that higher cognitive functions were the results of "mass action" of neurons, and were not susceptible to localization. (Carter, R., *Mapping the Mind*, Berkeley, CA: University of California Press, 2010.) However, direct electrical stimulation of the brain and a variety of scanning techniques such as nuclear magnetic resonance (NMR) or functional NMR that "lights up" areas of brain in "real time" that are actively functioning have demonstrated that certain areas of the brain are dedicated to specific functions. For example, speech and language can be mapped to Broca's and Wernicke's areas of the brain (we will return to this later under language). However, when we get to things like intelligence or consciousness, it has not been possible to point to a single area and localize a map of such entities. There are several possible explanations for this, including the probability that both intelligence and consciousness are "emergent phenomena" resulting from the function of many different areas of the brain. The other possibility is that we have a difficult time even defining both concepts, and the functional location (map) depends on what aspect of either intelligence or consciousness that we are considering.

The difficulty with satisfactorily understanding, defining or localizing intelligence within the brain makes it even more difficult to prophesize regarding future higher intelligence. However, we should not wrap intelligence in a veil of mystery. Steven Weinberg, the Nobel Prize winning physicist and author of multiple books both popular and scholarly on physics, has made what appears to be an appropriate comment on intelligence:

"We have not yet understood consciousness or intelligence, but there is no reason to suppose that these are anything but the workings of physics and chemistry within the brain. As far as we can tell, there is nothing in the fundamental laws of nature that will suggest any special role for life or intelligence in the plan of things." (Weinberg, Steven. "At Last--What Sort of World Is This?" *Forbes*, The Big Issue, 1996.)

David Wechsler, a famous psychologist, and early researcher in the area of intelligence, who also developed an intelligence test, defined intelligence as:

"The global capacity to act purposefully, to think rationally, and to deal effectively with the environment." (http://www.top-psychology.com/9041-Alfred%20Blnet/modern.asp)

We could look at many other definitions, but this one is probably as good a definition as any other. However, after reading the definition over several times, it is also obvious that the definition only deludes us into thinking that we have an understanding of what intelligence really is from the definition. The definition really tells us nothing useful that allows us to understand intelligence in any fundamental way. In addition, Wechsler's definition does not recognize the broad ranges of intelligences that we are interested in here.

In 1920, E. L. Thorndike divided intelligence into three types based on the ability to manage and understand: 1) ideas (abstract intelligence), 2) concrete objects (mechanical intelligence), and 3) people (social intelligence). (Kihlstrom, J. F., Cantor, N. "Social Intelligence." http://ist-socrates.berkeley.edu/~kihlstrm/social_intelligence.htm)

However, additional types or perhaps better characterized as facets of intelligence have now been recognized. For example, Premack and Woodruff in a now classic paper in 1978 introduced the term "Theory of Mind" from studies on chimpanzees. As the concept has been developed, Theory of Mind is the ability to understand or anticipate what others are thinking, and it is a form of intelligence not previously recognized. Theory of Mind was nicely described by Carl Zimmer in *Science* (taken here from Carl Zimmer's website: https://carlzimmer.com/how-the-mind-reads-other-minds-542/):

"Imagine a boy sitting on a couch about to unwrap a chocolate bar. His mother announces that it is time for her to take him to soccer practice. He tucks the chocolate under the couch for safekeeping and leaves. A few minutes later, his sister comes into the room in search of her teddy bear. When she looks under the couch. She is surprised to find an unopened chocolate bar, which she then decides to save, and hides it behind a bookshelf. When her brother comes home, drooling for chocolate, where will he look?

This may not seem like a difficult question: It's glaringly obvious that the boy will look under the couch. But to get the right answer, you have to perform an extraordinary mental feat: you must understand the boy's intentions and beliefs—regardless of their accuracy—and use that information to predict his action despite that you know that the candy bar is now behind the bookcase. And the skill doesn't come easily: Until the age of 5 or so, children will answer that the boy will look behind the bookshelf, where they know the chocolate bar to be."

In other words, the intelligence in this case is not in knowing where the chocolate bar is located, but in knowing where the boy thinks that the chocolate bar is located.

Interesting, like many other higher intelligential functions such as language and mathematical abilities, Theory of Mind appears to be unique to Humans and possibly (to a varying degree) to a few other primates. We are not born with the ability (Theory of Mind), but we develop it after about age 5, as we develop more advanced language and social behavior.

Theory of Mind, in addition to giving some insight to the workings of the mind, appears to also be involved in a neurodevelopmental disorder. Autism is a complex developmental disability that usually appears during the first three years of life and persists into adulthood. It is a relatively common disorder occurring in 0.2 to 0.6% of the population. Autistic individuals appear to have poor development in their Theory of Mind—they do poorly on the "chocolate bar test." In other words, they are limited in their ability to understand what someone else is thinking. Autism is characterized by difficulty in communication, poor social interaction and impaired creative or imaginative play as a child.

In addition to the Theory of Mind, the concepts of intelligence(s) have continued to expand. Arguably, one of the most widely accepted current concepts of intelligence is the theory of multiple intelligences suggested by Howard Gardner in 1983. Most concepts of intelligence in the past have centered on only a few aspects of intelligence, however, Gardner recognizes seven different types of intelligence that go well beyond the "classic" concepts. Gardner's concepts of multiple intelligences are listed in the table below.

Gardner's Multiple (7) Intelligences

Type	Characteristics
Logic-mathematics	Think logically, reason deductively, number manipulation
Linguistics	Syntax, ability to manipulate language, semantics
Spatial	Ability to create and manipulate mental images
Musical	Recognize and compose sound in terms of pitch, tones, and rhythm
Bodily-Kinesthetic	Ability to recognize and coordinate bodily movements
Interpersonal	Ability recognize the intentions and feelings of others
Intrapersonal	Ability to understand one's own feelings and motivations

Concepts from *Frames of Mind* by Howard Gardner, 1983

One of the important advancements in Gardner's classification of intelligence is that it begins to recognize intelligence in a broader perspective.

Some researchers in intelligence also believe that there exists a global factor that permeates all aspects of cognition, which is called Spearman's g factor. Factor analysis is used to statistically extract g from mental ability tests. (Gottfredson, L. S. "The General Intelligence Factor." Nov. 1998, http://www.psych.utoronto.ca/%7Ereingold/courses/intelligence/cache/1198gottfred.html)

The volume of frontal lobe grey matter, and the speed and reliability of neural transmission has also been correlated with intelligence and with the g factor. (Gray, J. R., Thompson, P.M. "Neurobiology of Intelligence: Science and Ethics." *Nature Reviews*, vol. 5, June 2004.)

However, all these approaches still tend to be based on very limited and restricted concepts of intelligence, and in addition are highly anthropocentric—for man only. Although it is true that man is the best (and the only) example of higher intelligence that does not (or should not) preclude the recognition of more limited forms of intelligence in "lower life forms." In other words, mankind is not the only creature with intelligence.

In addition, with genetic engineering it might be possible to produce increased intelligence even in other animals—like a dog or a horse. In fact, recently a smarter mouse has been bred which learns faster and remembers more—which can be considered a more intelligent mouse. This strain of mice is called "doogie" mice after the teenage genius Doogie Howser, M.D. depicted on TV. The mice were given extra copies of a gene called NR2B, which encodes a receptor for the neurotransmitter N-methyl-D-aspartate (NMDA), which has been shown to be involved in learning and memory. (Tang, Y., et al. *Nature*, vol. 401, Sept. 2, 1999, pp. 63-69.) These genetically engineered mice have memories 4 to 5 times as long as normal mice after insertion of this single gene into the mouse genome. This is an important concrete example of genetic engineering being used to make a small change resulting in an increased intelligence in an animal. In other words, this is an example of "proof of principle" that genetic engineering can be used to increase intelligence. It has already been done!

Although Gardner's seven intelligences are possibly valid as a first approximation, most do not readily lend themselves to discussions of "advanced intelligence" since they cannot be either easily scaled or readily generalized, nor is it clear how such intelligences could be used as basis for something "more." In other words, they do not readily lend themselves to new organizational concepts or increases in complexity which might yield an emergent advanced intelligence, which is what we seek. Although probably correct from their perspective, concepts of intelligence such as Gardner's or even Thorndike's simply do not help us "go further" in any discussion of advanced intelligence. We will therefore shortly look at intelligence from a different perspective—at a more basic level that will (hopefully) allow at least a small increase and more fruitful discussion of possible advanced intelligence. In addition, the bottom line is that we really do not know exactly what constitutes intelligence except in some specific forms such as the use of language and abstract manipulation of symbols and social interactions as epitomized by Gardner's concepts of seven forms of intelligences.

Measuring Intelligence

Most of us are familiar with the fact that Human intelligence is generally measured as an intelligence quotient or IQ score. Alfred Binet, a French psychologist, developed the first intelligence test in 1904. Binet was asked by the minister of education in France to develop a method of distinguishing "slow learners." Binet's original IQ test was essentially a vocabulary test. The "intelligence quotient (IQ)" was the mental age (as determined by testing) divided by the chronological age multiplied by 100. In other words, a seven-year-old that has the same average test score as other seven-year-olds will have a score of 100 (7/7 x 100 = 100 or average intelligence). However, a seven-year-old whose tests scores are at the level of the average ten-year-old will have an IQ of 143

(10/7 x 100 = 143). The value of this type of testing tends to level off at around age 16 and cannot be applied to adults. In order to be able to test and compare adults, a "deviation IQ" test was developed based on a bell-shaped curve (rather than a ratio as just described). With a "deviation IQ," intelligence is compared among individuals of the same age. For example, using one of the Wechsler IQ scales, the average Human intelligence is generally set at a score of 100 with a bell curve type distribution with a score ranging from 85 to 115 considered "normal"—one standard deviation which would include about 2/3 of the entire population.) (http://members.shaw.ca/delajara/IQBasics.html)

If we consider two standard deviations (on either side of the mean), that will give IQ scores of 70 to 130 (Wechsler IQ) and include all but about 5% of the Human population. Those below two standard deviations (SD) from the mean (IQ less than 70) are generally considered mentally challenged, and those above two SD (IQ greater than 130) are considered "bright." An IQ of 130 to 70 (+ or - 2 SD) includes 97.725% of the world's population.

Medically, mentally challenged individuals have in the past been classified as borderline (IQ 70–80), morons (IQ 50–69), imbecile (IQ 20–49), and idiots (below IQ of 20). Unfortunately, this terminology for mentally challenged individuals, although precisely defined, has been so socially abused that they are generally no longer used medically. Instead, mental retardation is now generally classed as mild (50–69), moderate (35–49), severe (20–34) or profound (less than 20).

Individuals with higher IQs (greater than 130) are often considered gifted. Only about 0.135% of the population has an IQ of 145 or greater (three standard deviations from the mean), and only three people per 100,000 population have an IQ of 160 or greater (four standard deviations from the mean). It is not exactly clear what is the highest IQ present in the world today. Somewhere around 7 SD with an IQ of 205, the cumulative percentile becomes so rare that such an intelligence probably does not exist in the world population of 7 billion people. In fact, the highest current IQ may be around 190 or possibly a bit higher.

Marilyn Vos Savant, who works as a columnist for Parade magazine claimed the highest IQ known. At age 10 she tested at a mental age of a 23-year-old, which gave her an IQ of 228—the highest score ever recorded. (Yam, P. "Intelligence Considered." *Scientific American Quarterly*, winter 1998.) However, using a "deviation IQ" (as described in the graph above) rather than the "ratio IQ," this high a score would probably not be possible. In addition, there is the problem of how do you test for such high IQs. It has been suggested that using the "deviation IQ," Marilyn's IQ would probably be about 190 or so (six standard deviations above the mean). Six standard deviations would include all of the population except 2 per billion. With a world population of 7 billion we would expect that there would be only 14 individuals with an IQ of 190 or more.

In some respects, IQ "testing" has literally become the de facto definition of intelligence. This is, of course, somewhat a form of circular reasoning—the test measures intelligence, but intelligence becomes what is being measured by the test. In Binet's original example, intelligence was, in essence, your vocabulary skills (in French). Although language (and therefore vocabulary) has played an important part—even a major part—in the development of Human intelligence, language is clearly not the only ingredient in the stew. However, having a large vocabulary does reflect some language skill, and requires some learning ability, some memory capabilities and perhaps at least some education (formal or informal), and those "intellectual skills" are therefore "buried" within the test in addition to the vocabulary per se. At least when viewed in this broader perspective, there may be some veneer of legitimacy to these early approaches using a vocabulary test for intelligence testing. However, vocabulary tests and many other aspects of IQ are linked only to Humans. This "linkage" inevitably affects and limits our basic concept of intelligence.

However, current tests measure (or attempt to measure) short-term memory, abstract reasoning, some vocabulary and mathematical skills, factual knowledge and information, visual-spatial abilities, and even comprehension and "common sense." Although IQ tests have come a long way, and there are even tests that are completely non-verbal (Leiter test) and can be used in individuals who do not speak a particular language, all tests still have some cultural bias, and it is still not clear that all aspects of intelligence are being adequately evaluated.

Although the concept of IQ testing has been refined over the years, and probably serves the limited purpose for which such tests were designed, such as evaluating scholastic capabilities or identifying the mentally challenged individual, this still leaves a great deal to be desired when we consider intelligence on a broader or absolute scale compared, for example, by the way we can understand and scale temperature (we will return to this shortly).

What is the "Hardware" for Intelligence?

The problem of "understanding" intelligence really lies in the same morass as understanding the mind and consciousness. There has been no lack of attempts to understand intelligence or the mind, however, we know (or think we know) that intelligence and the mind resides in the physical structure of the brain. The brain as a physical structure is much easier to understand than the more metaphysical concepts such as "mind" or "intelligence" or consciousness (whatever they are). The brain has been studied scientifically for years. Brain anatomy is well-known. The histology of the brain reveals structures at the cellular level, and collections of neurons called "nuclei" (unfortunately, the same name as the atomic nucleus, but these are collections of neurons often that perform a similar function) and their interconnection between themselves and the six neuron layers that constitute the cerebral cortex has also been intensely studied.

Neurons are the key cellular structures of the brain. They pass electrical signals, and act like biological wires and switches. The neuron itself consists of a cell with two sets of cytoplasmic extensions, the axon and dendrites. Signals come to the neuron through the dendrite, which branches extensively, and the "signals" are conducted away from the neuron by way of a long, thin extension—the axon. The connection between neurons is called a synapse. The neuron conducts an electrical signal, which results from depolarization of the cell membrane. Not every "signal" from one neuron's axon to another neuron's dendrite results in depolarization and the transmission of an electrical signal. That is why the neuron can act like a "switch" (transmitting an incoming signal or not transmitting the signal) as well as a conducting wire from an incoming signal to an outgoing signal (when a signal is transmitted). The neuron is quite talented. There are about 100 billion neurons in the Human brain. It is the arrangement and function of these neurons that are the basis of Human intelligence. The neuron has also been one of the driving forces of evolution, resulting in predator avoidance, pray acquisition and management of environmental threats and opportunities and the utilization of language and the ability to reason abstractly and do mathematics. It is all in the neurons and their connections. It is amazing that intelligence can emerge from this jumble of neurologic connections.

Myelinated axons (axons wrapped in an insulating material called myelin) can send signals 100 times faster than unmyelinated ones. Myelin also allows information to be sent by reducing the waiting time between signals. Myelinated neurons can therefore process 3,000 times as much information. (Emily Singer quoting George Bartzokis, *MIT Technical Review*, Nov/Dec 2009) This "anatomic change" (myelination) can therefore be viewed as increasing intelligence.

When the neuron is at rest, there is a negative electrical charge inside of the cell relative to the outside of the cell of about -70 millivolt. This is the resting potential. This negative charge is due to an excess of negatively charged proteins inside of the cell (with a smaller amount of negatively charged chloride ions (**Cl·**)) compared to positively charged potassium (**K·**) ions, and a much smaller amount of sodium ions (**Na·**). Outside the cell, the concentration of sodium (**Na·**) and chloride (**Cl**) are much larger.

However, the neuron cell membrane is packed with ion channels that are very specific for which type of ion they will allow to pass through the cell membrane.

When a "signal or impulse" is being passed, the sodium (**Na·**) channel opens, and positively charged sodium ions rushes from the outside into the cell and neutralize the overall negative charge inside the cell, and the cell membrane becomes positively charged. The large negatively charged protein ions cannot easily cross the cell membrane and remain inside the cell. However, within a millisecond, the sodium channel closes, and the potassium channel opens, letting (**K·**) ions leave the cell again creating a negative charge inside the cell compared to the outside. This results in a depolarization "spike" that moves along the cell membrane of the neuron. This is somewhat like an electric charge moving through a wire.

An action potential or depolarization spike resulting initially from the selective entrance of sodium ions (**Na·**), is followed by re-establishment of the negative charge following the exit of potassium ions (**K·**).

Although a beautiful bit of simple electrochemistry based on ion "gates" resulting in a conducted electrical signal—the action potential, it would only work once. There needs to be a mechanism to re-establish the sodium (**Na·**), and potassium (**K·**) gradient that existed before the depolarization. Mother Nature is very clever—there is an active "sodium pump" (that means it requires energy to work) that pushes sodium back out of the cell after an action potential. For every three sodium ions (**Na·**) actively pumped out, two potassium ions (**K·**) are let back into the cell, re-establishing the original ion concentrations across the neuron cell membrane.

The action potential is "all or nothing", and the depolarization (and therefore the "signal") moves in only one direction: from the dendrites through the axon.

The actual connection between neurons is at the synapse. The synapse is the junction between neurons where information can be modulated. The synapse is where the "switch-like" action of the neuron occurs. The arrival of an action potential at the synapse results in the fusion of some vesicles with the cell membrane and the release of chemicals called neurotransmitters. The release of the neurotransmitter results in an excitatory postsynaptic potential across the synapse. This may or may not result in an action potential and transmission of the signal. The principal excitatory neurotransmitter is glutamate. Glutamate is an amino acid, and in addition to serving as a neurotransmitter at synaptic junctions, it also serves as one of the building blocks for peptides and proteins. In addition to excitatory neurotransmitters, there are also inhibitory neurotransmitters. GABA (gamma-aminobutyric acid) is the major inhibitory neurotransmitter. GABA can prevent or inhibit the transmission of an action potential. There are also a number of other modulators that act at synaptic junctions like acetylcholine, dopamine, epinephrine, and others. There have been 50 different neurotransmitters identified (Carter, R. *Mapping the Mind*, University of California Press, 1998.)

Why is any of this important? This very basic quick review of the neuron is simply intended to help demystify these very basic components of the brain—the neuron and the synapse and note that the mind and intelligence emerges from the complex interaction of these basic building blocks. The neuron and the synapses connecting neurons are marvelous inventions of evolution, and they are the basis of everything we think and know. However, the system is neither mysterious nor magical. Nevertheless, any advance in Human intelligence will require a scaling or perturbation of this system. In addition, one might ask; "where is the intelligence in all of this?" The intelligence is at a higher level of organization…in the circuitry and connections in the brain, which we will look at (generally indirectly) as we review the components of intelligence. However, the important point is exactly as Weinberg noted: **"We have not yet fully understood consciousness or intelligence, but there is no reason to suppose that these are anything but the workings of physics and chemistry within the brain."**

Components of Intelligence

Although there is no defined consensus as to exactly what constitutes intelligence, we can probably make some useful progress in the understanding of the concept if we broaden the discussion to some of the components of intelligence. Components are simply the characteristics that pretty much everyone associates with intelligence. However, the components of intelligence are clearly not intelligence by themselves. We can start with the usual suspects, and then expand into some components that are not as clearly currently recognized as aspects of intelligence. However, again it should be kept in mind that none of the components that we will be discussing are not intelligence individually by themselves. Intelligence is more than just the sum of its parts. (we have seen a similar phenomenon in life itself). The "new buzzword" for this type of phenomenon is "emergent"—in essence, more appears to come out than was put in. A tire or a piston is not an automobile. They are components of an automobile. However, when the right parts (components, if you will) are put together in the appropriate manner, an automobile is created—you might say that it is an emergent phenomenon; an automobile becomes more than the sum of its components. You can drive around in an automobile, but you cannot drive around in a pile of automobile components heaped on the ground. We have looked at this metaphor previously when we considered the development of life. In other words, some combinations of components can result in a "value added" which conveys a new level of function or a new level of "something" which was not present before.

In addition, it should become clear as we look at our laundry list that they are not all necessarily "pure" components. Some of the components are obviously made of parts of other components. In other words, the components blend together. It is sometimes difficult to separate where one component ends, and another begins. However, despite the limitations of dissecting out and working with the components of intelligence rather than trying to work with intelligence per se, I would opine that this approach (components) will be the boat (vehicle) to get us across the river on this part of our journey.

What we are doing is following Rene Descartes' advice: "Divide each difficulty into as many parts as is feasible and necessary to resolve it." (http://en.wikiquote.org/wiki/Ren%C3%A9_Descartes)

After breaking intelligence down into **components**, we will examine the prospects that such components can be scaled (simply add more), generalized (expand the component into potential new realms), or organized with increased complexity to potentially develop emergent properties of advanced intelligence (stir the stew). In addition, implicit in the "component approach" is that it is simpler to recognize the components of intelligence in a "lower" animal. In fact, we should also be able to readily recognize at least some intelligent components in non-biological systems such as computers or even aliens (if we ever find one). The presence of one or more components of intelligence does not necessarily guarantee the presence of intelligence per se, but the presence of several components does at least begin to raise the question of where the components end, and where something that we would recognize or define as intelligence would begin. If we stir enough components into the stew, do we have intelligence? That is not entirely clear at this point in time. We would probably have to add one or perhaps two other ingredients—consciousness and possibly even the mysterious g factor, but we will "add" them in a slightly different manner than the other components as you will see.

You will probably immediately recognize that we are throwing anthropocentric intelligence out of the window—it is simply too limited and restricted of a concept of intelligence to be useful to us here in a discussion of future higher intelligence. Implicit in this approach is that mankind has intelligence, but we do not have an exclusive contract, and it should be possible to increase mankind's intelligence in the future. I do not want to lose the concept that Human intelligence has advanced to the point that we have another "value-added" emergent phenomena, which I will refer to as a sentient being. I have already conjectured (argued) that this is the first step toward Godship (however, I have also noted that we [Humans] have the minimal amount of whatever this "value-added emergent phenomena" or entity is). However, whatever it is; it is not supernatural, and can be found in the complexity of the Human brain.

What are the components of intelligence? Actually, they are not at all mysterious…they are rather obvious. You already know them. We will start out lumping our components into ten large component categories:

1. **Input**
2. **Processing**
3. **Anatomy and Functional Parameters** (the hardware that gives the reality to the rest)
4. **Short and Long-Term Memory**
5. **Learning**
6. **Problem-Solving, Deductive Reasoning,** and **Abstract Symbol Manipulation** (including mathematics and logic)
7. **Language**
8. **Knowledge Base**
9. **Emotions (including social intelligence and Theory of Mind)**
10. **Output**

In addition, there appears to be one additional factor—**Consciousness (and possibly Spearman's g)**. I am setting consciousness aside and will be treating it somewhat differently than the other 10 components because we really do not fully understand it at this time, and I am going to be using it as a multiplier of intelligence as I will describe shortly.

Consciousness is as slippery a concept to get a hold on as intelligence itself. Nonetheless, we will consider consciousness as a factor of intelligence, but as in the *King and I* movie, we will love it more, but trust it less. As we will shortly see, consciousness and Spearman's g will be multipliers of our other summed components (but I am getting ahead of my story, and we will return to this concept later). When possible, we want to track our components in a classic reductionist approach all the way down to the brain anatomy and neurons. Please note that we are trying to understand consciousness and intelligence rather than give a scientific definition.

If some of these components (like processing) sound like computer jargon…they are, but they are also buried in the neuroanatomy and physiology of the brain, and they will be useful in dissecting out intelligence and trying to "engineer" a future more intelligent Human.

So where is intelligence? Intelligence is not at the level of the organization of the components. Intelligence is at the level of organization of the emergent phenomena that comes out of the brain, after we stick all of the components back together.

However, we will also not delude you further (since I am sure many will immediately recognize or at least suspect where we are going). As we have alluded, we are going to break intelligence down into some components and try to envision how these components could be scaled, enhanced, generalized, or otherwise perturbed to produce an advanced intelligence or at least a first approximation. The possibility of scaling, enhancing, generalizing, or perturbing the components to arrive at a more advanced intelligence requires that we understand the anatomic and neurological basis of these components. Scaling will be the obvious easiest method of enhancing intelligence since it simply means that we genetically engineer more of what we already have. Generalizing or perturbing a component will be more difficult and will probably only be recognized for most components as more experimental brain research is done. However, in the end we will put the components back together and consider intelligence as the emergent phenomena resulting from the complex interaction of the summed composite components multiplied by this thing called consciousness. We will also leave the door open for additional components, and perhaps one additional multiplier.

But where are the advanced intrapersonal intelligence, interpersonal or social intelligence, Theory of Mind or any of Gardner's concepts of multiple intelligences? With components, we are clearly looking at a different approach and sometimes different level of organization to the types of approaches to intelligence used by Gardner. In fact, most of the types of intelligence described by Gardner involve many or even most of the components of intelligence that we will be discussing. We might say that Gardner's concepts of intelligence are a composite of various components (with the probable exception of bodily kinesthetics, which Gardner raises to a level of one of his seven intelligences, and we will be considering as a part of input functions (discussed next). Also, music would be an input function coupled with some processing, memory and emotional thinking, and perhaps output (dancing or tapping one's foot).

Input

"**Input**" may initially sound like a strange component of intelligence. We test for it on IQ tests (but not directly). Yet, input is in fact, an integral component of intelligence. All information that the brain receives from the outside world is input. A great deal of our input comes from our five senses, which were originally described by Aristotle in ancient Greece—sight, hearing, smell, taste, and touch. However, there are at least 15 other senses and possibly even more depending on how one defines "senses," and whether one is a "lumper" or a "splitter." For example, the sensation of weight, position of the body and the amount of bending of the various joints really provides different forms of information, but they are generally lumped together as proprioception. In fact, Gardner, as noted above, elevates this group of "body awareness" sensory inputs to the level of one of his seven forms of intelligence. Despite its obvious importance, we will allocate proprioception to one form of input, which is only one of our components of intelligence.

Other sensory inputs include the sense of balance or equilibrium, orientation, and the sense of starting or stopping that originates in specialized sense organs of the ears (in addition to hearing). There are also general senses such as hunger, thirst, and fatigue. The sense of pain can be separated into cutaneous (skin), visceral (body organs) and somatic (muscle, joints, bones) pain. In addition, the sense of touch really comprises many assorted forms of input including vibration, pressure, stretch, itch, texture, and temperature.

However, to really "see" sensory input as a vital component of intelligence, we can do a simple "thought experiment," which was made famous by Einstein. Though experiments are generally experiments that could never be done in the laboratory or anywhere else for that matter—that is why they are relegated to being an experiment only in the mind. In our experiment, imagine a Human brain—let's use Einstein's brain (surely, he would not mind us borrowing it for our experiment, since it is a type of experiment which he invented). We will isolate Einstein's brain from inception (even before birth) and cut it off from all sensory input of any kind whatsoever…nothing gets **into** Einstein's brain…nothing. Then we ask the question: "What would Einstein know about the world at age 20?"

"…absolutely nothing."

Would he develop language by himself? The answer is surely not. In fact, there is evidence that nerve tracts that are not used after birth degenerate. With no input from the outside world, how intelligent would Einstein be? If Einstein received no input of any form…no light, no sound, not even any proprioception, then what would he know of the world? Would he even be able to figure out that there is an "outside world?" Remember, we are restricting **all** input from inception. Einstein cannot feel a chair or a table or anything, therefore he does not even know that there is such a thing as an object outside his brain. In fact, since we are restricting all input such as touch or proprioception, he would probably not even know that he exists. Since he could not sense the effects of gravity, he would not know what is up or down. He would not even be able to figure out whether there is a concept of up or down. What our thought experiment illustrates is how important input is to intelligence. No input, literally no intelligence! A Human brain that is completely isolated would not be able to think.

One implication is that increasing input would be one approach to increasing intelligence. However, input is intimately integrated with other components such as learning, processing, and memory.

Are there other types of input that Humans do not have that could improve intelligence?

In the component approach to intelligence that we are using here, any increase capability in any of the components would be translated into an increase in intelligence. However, the increase in intelligence resulting from some types of increased input might be trivial.

For example, we are discussing input, so let's take the example of hearing. Human hearing ranges over about 20 to 20,000 cycles per second (Hertz). Yet, dogs can hear over a range of about 40 to 46,000 Hertz, and horses can hear between 31 and 40,000 Hertz—clearly the capacity to hear over a much wider range of frequencies than Humans can hear is well established in some biological systems. If we genetically manipulated the genes involved in Human hearing to make them comparable to a dog, would we be more intelligent? Clearly, we would have better hearing in the higher frequency ranges (increased input), but would that make us more intelligent? Using our concept here that any increase in the components of intelligence would be an increase in overall intelligence the answer would have to be yes. However, in this case the increase in intelligence would probably be small to minimal.

Input is not measured **directly** on current IQ tests, but even if input components such as hearing were a part of our concept of IQ testing, the increase in intelligence resulting from a greater range of auditory frequencies would probably be much less than one IQ point. Likewise increased hearing sensitivity might be a nice addition, and possible, even easy to do genetically, but it would probably "add" very little in terms of advancing intelligence. In other words, if we were to try to assign an increased IQ resulting for increasing the frequency range of hearing we might be talking about 0.01 …in other words a trivial increase.

We might look at vision in a similar manner. Actually, we see only a very small fraction of the electromagnetic spectrum—visible light, which has a wavelength of between 7×10^{-5} to 4×10^{-5} cm. The eye is a wonderful invention of evolution, and vision is one of our most important senses in terms of information input regarding the world in which we live. However, visible light forms only a very small portion of the electromagnetic spectrum. Could we increase the electromagnetic spectrum that we visualize? Would that increase our intelligence?

In other words, we are looking through a pinhole at the electromagnetic spectrum, which extends from long wavelengths such as radio waves that can be kilometers long, to microwaves, to infrared, to light. Shorter wavelengths of light include ultraviolet, x-rays and gamma rays.

We, of course, are already capable of "seeing" the entire electromagnetic spectrum, however, we have to use different instruments to see different parts of the whole spectrum. For example, we use a radio receiver and antenna to hear music (which is a compression wave and not an electromagnetic wave) that was converted into a radio wave (which is an electromagnetic wave) and is broadcasted, and the radio wave is then converted back to a compression wave by a receiver with a speaker…and we hear it. Different instruments are used to intercept the electromagnetic radiation and "translate" it into a form that we can understand (like a meter reading the intensity and/or direction).

Would we be more intelligent if we could "see" the whole electromagnetic spectrum? Using the "ground rules" we are operating under, the answer would again have to be yes, but again it is hard to imagine how this would make a major difference in intelligence other than what we are currently doing. We are already "seeing" most of the electromagnetic spectrum (EMS) using instruments of one form or other, and it is not clear if seeing the whole EMS with our eyes would add significantly to our intelligence. However, for most of mankind's existence, we did not have radio available. In addition, since there are only a few examples in nature such as insects seeing in the ultraviolet range, it is not easy to envision (pun intended) where we would get the genetics to accomplish this wider range of "vision" of the entire electromagnetic spectrum. Since it does not currently exist in biology, we would literally have to genetically engineer it from "scratch." That type of genetic engineering may someday be possible; however, it is probably so far in the future as to make even speculation meaningless at this time. However, radio (and probably television) have, in fact, increased mankind's intelligence, but putting a number on our IQ because of their use would be difficult.

Increasing most of our current senses in terms of their range or sensitivity would (by the consideration used here) increase our intelligence, but again most of the increases would probably be trivial unless we could utilize it in some form for more rapid learning or "advanced" communication.

In fact, if we look at the Human brain anatomy, a very large part of the brain, including the cerebral cortex, is devoted to sensing our environment with sight, hearing, proprioception and our other "inputs."

What is really needed in the area of input is the ability to get large amounts of data or information into the brain rapidly…something like downloading a book-size file from the internet in a few seconds. However, it is not clear how that could be done without resorting to some bionic capability that is currently science fiction. We will speculate on one approach in the next chapter (*The Dumb Machine*) when we discuss neural interfacing. We will be looking at the possibility of inputting data directly into the brain electronically rather than through any of our current senses. Although increasing input has the potential for dramatically increasing intelligence, pragmatically, input will probably continue through our current senses, such as vision and hearing. However, all is not lost. Since input is intimately integrated with processing, learning and memory, it may turn out that our current input such as vision is being underutilized. In other words, the limitation on data download (such as reading a book) may not be an input restriction, but a restriction at the level of processing, learning, memory, abstract symbol manipulation and language. In other words, if we can scale up all of these other components, we may find that our current "inputs" like vision and hearing are completely adequate even at their current level of function. If we can read a book, understand it and remember everything we have read as fast as we can physically turn the page, then clearly our current vision is completely adequate and compatible for input in a being with much higher intelligence.

Although some small increase in intelligence may be possible by expanding the range of our current senses, probably the greatest contribution to increasing intelligence will be with increasing the **rate of input,** and even developing the possibility of a new direct electronic approach to putting information, experience, etc. directly into the brain.

Processing

Processing is basically the manipulation of data. Computers manipulate data, and the Human brain also manipulates data. Of course, the Human brain and computers do this by entirely different mechanisms. There have been a number of attempts to express the function of the Human brain in terms of computer (engineering) functions to allow comparison and to get at least a "ballpark" estimate of how much processing power that the Human brain contains. Ray Kurzweil in his book *The Age of Spiritual Machines: When Computers Exceed Human Intelligence*, and others like Nick Bostrom have made at least reasonable first attempts at approximating the Human brain's processing capability.

"The Human brain contains about 10^{11} neurons (most neuroscientists agree on this estimate). Each neuron has about $1-10 \times 10^3$ synapses, and signals are transmitted along these synapses at an average frequency of about 10^2 Hz. Each signal contains perhaps 5 bits. This equals to 10^{17} operations per second (OPS)." (Bostrom, N. "How Long Before Superintelligence?" *Kurzweil.net* April 30, 2001, http://www.kurzweilan.net/articles/art0168.html)

Expressing brain function as OPS allows comparison to computer processing functions. This estimate equates Human brain function with about one hundred thousand trillion computer operations per second (10^{17} OPS). One of the main factors contributing to this very high estimate of Human mental ability is massive parallel processing by the brain. However, as recognized by both Bostrom and Kurzweil, the 10^{17} OPS is not really realistic, and is best considered a very upper limit (if possible), because of redundancy in the neural processing which they estimate might reduce the Human brain processing capability (in computer terms) to about 10^{14} OPS.

This Human brain processing capability can then be roughly compared to the processing power of supercomputers, which is generally measured in FLOPS or the number of floating-point operations per second. In essence, FLOPS is a measure of how fast a computer system can handle floating-point numbers, and it is one of the industrial standards to compare computer performance or "power." A floating-point operation includes any operation that involves fractional numbers. These operations take longer than operations involving integers. On December 17, 1996, the Department of Energy's Sandia National Laboratory and the Intel Corporation announced that they had broken the teraflops barrier (a trillion floating point operations per second) with 1.06×10^{12} FLOPS. This beat the previous performance record of 368.2 billion FLOPS by over 250%. (http://www.intel.com/pressroom/archive/releases/cn121796.htm)

The Department of Energy has spent $200 million dollars for the Advanced Strategic Computing Initiative (ASCI) computer to help study the effects of aging on nuclear weapons. This supercomputer performs 30 teraflops, and is named "ASCI Q."

However, the race has continued at a furious pace. The United States had lost the lead to China's Tianhe-1A supercomputer and Japan's Fujitsu K in 2009 but recovered the lead in 2012 with the Sequoia BlueGene/Q at Lawrence Livermore National Laboratory with 16.33 petaflops per second (peta = 10^{15} or 1 million billion). However, being number one in computation speed does not last long. The Titan Cray XK47 at Oak Ridge National Laboratory has reported a performance of 17.59 petaflops per second. In addition, storage arrays continue to increase. In January 2012, Cray began construction of the Blue Water Supercomputer which has a storage capacity of "more than 25 petabytes disk storage, enough to store all of the printed documents in all of the world's libraries; and up to 500 petabytes of tape storage, enough to store 10% of all of the words spoken in the existence of Humankind." (National Center for Supercomputing Applications,

http://www.ncsa.illinois.edu/enabling/bluewaters)

Shortly after the creation of the Blue Water supercomputer, China again took the lead in supercomputing and maintained that lead for about five years. (https://www.wired.com/story/worlds-fastest-supercomputer-breaks-ai-record/) In 2018, the world's current leading supercomputer named Summit (the 200 petaflop IBM AC922 supercomputer at Oak Ridge National Laboratory with 9000 IBM Power9 CPUs) achieved a speed of a billion billion (10^{18}) OPS, which is called an exaflop. Summit's incredible computing capacity is being used to study climate change, which is what the computer was analyzing when it achieved the record-breaking exaflop speed. Summit also works in other areas such as scientific queries in health, engineering, physics, and other areas. (https://www.olcf.ornl.gov/summit/)

The significance of these incredible accomplishments can perhaps be appreciated by noting the fact that Intel had introduced the world's first microprocessor that produced only 60,000 OPS in 1971. (http://www.intel.com/pressroom/archive/releases/cn121796.htm) That is an increase of twelve orders of magnitude in computer performance or about a tenfold increase about every 3.6 years over the next 25 years (up to 1996). This incredible rapid rate of expansion in processing speed even exceeds Moore's Law, and it is not slowing down. Computer speeds have continued to increase to what has been estimated now as being an eleven-fold increase every year with speeds doubling about every 3.4 months since 2012 when the Blue Water Supercomputer was created. (https://www.wired.com/story/worlds-fastest-supercomputer-breaks-ai-record/) Computer performance (as expressed in flops) and storage capacity is a rapidly upward moving target. The significance of this is that even the 17.59 petaflops (17.59 x 10^{15} flops) of the Blue Water supercomputer exceeds the computing power estimated by Kurzweil for the Human brain (10^{14} OPS) by two orders of magnitude.

Despite the fact that computer processing is now in the "ballpark" of the Human brain (or even exceeds the Human brain), no computer has yet called and asked for a salary raise or performed any of the shenanigans of Hal the computer in the movie 2001. Alan Turing had set up the famous criteria to judge when computers had reached Human intelligence. Turing's defined a simple criterion for comparing computer to Human intelligence: when a Human can no longer tell when they are talking to a computer versus another Human, then a computer has achieved Human intelligence (according to Turing). However, I personally think that a better criterion would be when a computer can think independently and take actions not controlled by Humans such as Clarke's Hal (a computer) in the movie *2001: A Space Odyssey* from 1968 when Hal began to act independently of Humans.

Of course, the Human brain does not operate with FLOPS or even OPS and the estimate of 10^{18} operations per second or even more conservative (and realistic) estimate of 10^{14} OPS for the Human brain is a gross oversimplification of Human thinking, and probably a considerable overestimate of any comparison of the Human brain to computer processing ability, especially when applied to intelligence. A considerable amount of the Human brain neuron capacity is devoted to "housekeeping" functions such as the innervation of muscles, regulation of breathing, etc. Although this can be legitimately considered processing and makes a major contribution to biological function, only a small part of the Human cortex (and an even smaller portion of the rest of the brain) is really involved in what might be considered intellectual functions. On the other hand, how much of computer processing can be considered intelligence. Human and computer "thinking" is different, and we must be careful when making comparisons.

Therefore, considerable care must be taken in utilizing any such comparisons. However, these comparisons may be useful, provided that we recognize that they are "ballpark" estimates (the ball is in the ballpark, but do not expect it to be at home plate). Biological thinking and intelligence are quite different from computer intelligence, and it is more than a "numbers game."

Processing takes place largely at the cellular level and involves the number of neurons, the number of synapses, the arrangement of synaptic circuits, the types of neurotransmitters and the rate of neuronal transmission and rate of repeat transmissions.

The fact remains that computer "power" appears to be approaching the "ballpark" estimates of the processing power of the Human brain, and considering many of the **components** of intelligence, some of the components of computer intelligence are already much more advanced than Humans. However, we need to emphasize again that computers lack a key component of intelligence which we call consciousness, which Human intelligence has not been smart enough to understand and reproduce "in silico" …so far.

From the perspective of the components of Human intelligence, the Human brain is an incredible parallel processor, and this component of intelligence should be very scalable.

We have previously noted that myelin sheaths increase the speed of neural transmission, and processing by a factor of 3,000. However, that has already been done by "Mother Nature," and we need to consider what is the "next step" that might advance Human processing? Perhaps the "next step" will involve increases in parallel processing or more neurons or more connections, which brings us to our next component of intelligence.

Anatomic or Functional Scaling

Anatomic or functional scaling is like looking at increasing the hardware of a computer. Will it be possible in the future to increase the "hardware" of the Human brain?

Giving more neurons and more synapses (that are not dedicated to housekeeping functions), might potentially produce more processing power and give more intelligence. In fact, "getting there" (more neurons) may not be nearly as complicated as one might think. We will look at that problem in the mouse shortly.

Perhaps the simplest answer to increasing man's intelligence and giving us more "processing power" is merely to increase our brain size—problem solved…maybe.

However, when we look around at brain sizes and intelligence, an immediate problem is noted. The animal with the largest brain size in the world is the sperm whale with an average brain size of 17 to 19 pounds. That is definitely a whopper in the brain department, and it is 6 to 7 times as large as the Human 2.5 to 3 pound mini-mind (mini at least in comparison to the whale). Yet, there is no argument that the sperm whale's large brain does not translate into 6 to 7 times greater intelligence than man. The problem is that the whale's large muscle mass and body size requires an enormous number of neurons (brain cells) to simply control all of its massive bulk. A large brain used simply to control a large mass of muscles does not translate directly into more intelligence. Obviously, body size is an important factor when considering brain size and intelligence. For example, Humans have a brain to body weight ratio of roughly 1 to 50, while the sperm whale has a ratio of about 1 to 4400. By that comparison man has much more brains per pound of body than the whale. However, that also cannot be translated directly into intelligence either. The sparrow has a brain to weight ratio of 1 to 20, and sparrows are unmistakably not intellectual giants—despite having considerable more brains per unit of body weight than Humans. (Marable, K. "The Neurological and Environmental Basis for Differing Intelligences: A Comparison of Primate and Cetacean Mentality." http://www.msu.edu/user/marablek/whal-int.htm) In fact, in the small worm, C. elegans, about one out of every three cells is a nerve cell giving a "brain" to weight ratio of about 1 to 3. Clearly, neither the weight of the brain without consideration of the size of the animal nor the brain to weight ratio (which does consider the size of the animal in terms of weight) can be used as a reliable guide to predict intelligence. There appears to be more involved. One ratio that man does finally win is the brain weight to body length. Humans have about 240 grams of brain per foot of body length. Dolphins are second with about 200 grams per foot. However, as pointed out by Marable, this is a rather arbitrary ratio. (Marable, K. "The Neurological and Environmental Basis for Differing Intelligences: A Comparison of Primate and Cetacean Mentality." http://www.msu.edu/user/marablek/whal-int.htm). Brain size does roughly correlate with intelligence, but only when the size and basic morphology of the animal is also considered.

In addition, there is only a modest correlation within a species between brain size and intelligence. MRI-based studies in Humans have noted a moderate correlation between brain size and intelligence of 0.40 to 0.51. (Gray, J.R., Thompson, P.M. "Neurobiology of Intelligence: Science and Ethics." *Nature Reviews/ Neuroscience*, vol. 5, June 2004.) **However, there are small-brained people who are very bright, and large-brained people who are not so smart. In fact, Albert Einstein, who would generally be included in the group of brighter Humans, had a brain that weighed only 1,230 grams, which is below the average Human brain weight of about 1350 to 1400 grams.** (*Neuroscience Letters*, vol. 210, 1996, pp. 161–164.)

Rather than simply considering brain weight and animal size when considering higher intelligence, it is probably more important to consider parts of the brain that are known to be involved in intelligent functions like the cerebral cortex. In this comparison, the Human cortical surface area is about 1,000 times as large as the mouse. However, the average mouse weighs about 50 grams, and the average Human weighs 70 kilograms (70,000 grams) (North American males have an average weight of 78 kilograms, and females have a weight of 62 kilograms). This translates into Humans weighing 1,400 times the weight of the mouse. Therefore, comparing cortical size per body weight, the mouse has more cortex per gram of body weight than a Human. Again, when we factor in weight, the Human cortex does not look so large.

However, there is a genetic "disease" called microcephaly in which the brain is supposedly perfectly normal in organization, but 3 to 4 times smaller than normal. We have previously discussed the genes responsible for microcephaly when we were considering the evolutionary development of mankind. This perfectly normal, but smaller Human brain generally results in mental retardation. This would strongly imply that brain size does matter…at least on the lower side of the size scale.

Still, if we are trying to increase intelligence by increasing processing potential, the important part of the brain to increase would probably be the cerebral cortex. How do you increase the size of the cortex? Two possibilities immediately come to mind. The first would simply involve scaling. This would involve an expansion of the current cortex. This might be done by increasing the folding of the cortex to pack more cortex into the same area. Increasing the number of cortical cells might be done by preventing cortical precursor cells from differentiating, giving one or two more cell divisions before becoming neurons. We are going to see how this was done with dramatic results in the mouse shortly.

Another method of potentially increasing processing potential of the cortex would be to do what evolution has done at least twice in the past…add a new layer of neurons or cortex on top of the current cortex. This approach might give the greatest potential for developing a really advanced intelligence. This "advanced cortex" would be engineered to fit over the current Human neocortex. However, since we do not fully understand Human intelligence, we are speculating on how intelligence could be increased. It is far from clear if simply giving the brain more neurons would increase intelligence.

During evolution, neuroscientists can recognize at least two major changes in brain anatomy, and each has literally involved putting additional neural tissue on top of an older "brain." This added brain tissue (neurons and connecting synapses through axons and dendrites) has in each instance resulted in a major advancement in intelligence (slow-witted reptiles evolved to more adept mammals to smarter primates to intelligent Humans. Mother Nature's way of increasing intelligence through evolution has been by literally sticking new neural tissue—in essence a "new" brain or neural cells—on top of the old one. On top of the reptilian brain (the R-complex); mammals developed the old mammalian brain (sometimes referred to as the limbic system). On top of the old mammalian brain developed the primate neocortex. However, any additional expansion would almost invariably result in a much larger head…something like the science fiction movie of the intelligent alien with a very large disproportionate head. That is not exactly something that most Humans would find aesthetically pleasing. Perhaps adding only, a new thin cortical area a millimeter thick that connected mainly to the frontal cortex would be acceptable. However, this is completely speculative, and the "science" of experimental brain research that is going to be needed to develop a second cortex has not even been invented yet.

Expanding the neocortex by adding all those wrinkles on the Human brain surface called gyri (ridges) and sulci (valleys) appears to have been an important factor, which resulted in a considerable enhancement in intelligence. In other words, the last dramatic development involved increasing the size of the cortex rather than adding a new layer of cortex. Why not just increase the number of gyri and sulci to pack more neurons into an even smaller space? Any significant increase may not be possible simply because the number of neurons and their connections to other neurons through axons and dendrites are already densely packed.

What would be the effect of doubling or tripling the size of the current Human neocortex? Clearly, we do not know. This scenario has something of the flavor of the movie *Field of Dreams* with Kevin Costner. In the movie, Costner plows under his crop of corn to build a baseball stadium under the influence of a voice telling him, "If you build it, they will come." If we add more neural tissue "on top" of what Humans already have, what would be the results? Clearly no one knows the answer, and it is almost impossible to even make an educated guess. We might look at the process of increasing neurological development (intelligence) from reptiles to mammals to primates to Humans in terms of emergent phenomenon, which we might characterize as the presence of something totally new and different, which could not be predicted from simply looking at what was added. What the emergent intelligence would be is beyond conjecture. In addition, there is also the not so small problem of "Mother Nature's" rather crude developmental techniques. We can be sure that over the millions of years that it took to "progress" from the reptilian brain to the Human brain, the path is literally filled with the corpses of failed experiments. "Mother Nature" has never shown herself to be in the least bit squeamish about the sacrifice of life or limb…Human or otherwise. In fact, as we have previously noted, the "process" of evolution continues today. Evolution is a disaster on the level of the individual, because almost all the "experiments" result in disease or death. At least in the future using **IDE**$_{tb}$, we can expect the development to be much more rational, and much less of a disaster than what evolution has given us in the past.

What is the possibility of realistically developing additional brain tissue—particularly by expanding the cerebral cortex by genetic engineering? Any real expansion of the brain might appear quite speculative—science fiction.

This possibility is not as "far-fetched" as one might think. Chenn and Walsh generated transgenic mice to over express a modified protein (β-catenin) by neuroepithelial precursor cells in the brain. The modified β-catenin stimulates neural precursor cells to proliferate (divide) more times before differentiating and forming the cerebral cortex. Transgenic embryos developed grossly enlarged brains with a considerable increase in surface area of the cerebral cortex. This increase in size resulted in the formation of folds of the normally smooth mouse cerebral cortex resembling the gyri and sulci noted in higher mammals (such as Humans). (Chenn, A., Walsh, C. A. "Regulation of Cerebral Cortical Size by Control of Cell Cycle Exit in Neural Precursors." *Science*, vol. 297, July 19, 2002.)

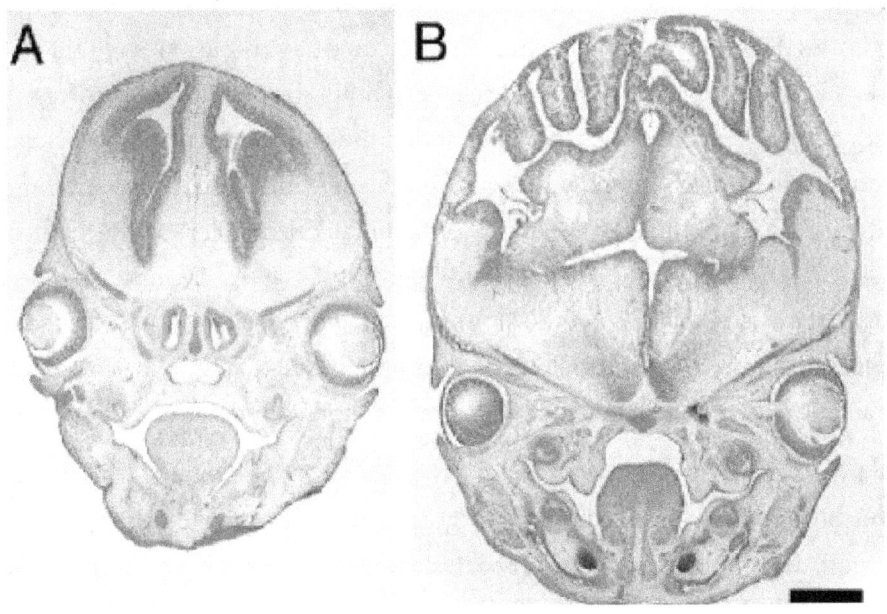

Brain and head of a normal mouse (A) compared to a mouse genetically engineered to express a modified protein called β-catenin (B). The figure shows the brains and head of a normal mouse (A) compared to a mouse with the genetically engineered modified β-catenin (B). Note not only the larger brain of (B), but also the folds (gyri and sulci) on the surface of (B)

compared to the smooth cerebral surface of the normal mouse (A). The "spontaneous" folding pattern in this mouse suggests that perhaps a similar mechanism has resulted in the cerebral folding pattern noted in Humans.

(Chenn, A., Walsh, C. A. "Regulation of Cerebral Cortical Size by Control of Cell Cycle Exit in Neural Precursors." *Science*, vol. 297, July 19, 2002.)

This rather interesting and elegant experiment suggests some important implications. The marked increase in brain size was accomplished by modifying only **one** protein. There has been considerable discussion about the development of the enlarging brain over millions of years from Australopithecus to Homo habilis to Homo erectus to Homo sapiens. Chenn and Walsh's experiment indicates that dramatic changes in brain size could have occurred very rapidly—even in a single generation. A rapid change in brain size would also be consistent with current genetic studies that suggest that the Human population developed from a rather small group. However, this does not imply that a mutation in β-catenin was responsible for the increase in Human cerebral cortex development. As we have previously discussed, other genes such as ASPM have also been suggested as possible candidates. However, this study does clearly illustrate that it does not necessarily take millions of years or hundreds of mutations to produce dramatic anatomical changes in brain size or in the size of the cortex.

Dramatic changes in Human brain size could probably be produced even now—possibly using the modified β-catenin system. However, it would clearly require a great deal more study, and probably further modification of the β-catenin system to **reduce** the magnitude of the cerebral expansion. Chenn and Walsh reported that the modified β-catenin (which had been NH_2-terminally truncated (the end of the protein chain was cut off, which made the protein less effective) resulted in about a twofold increase in the proportion of precursor cells that re-entered the cell cycle instead of differentiating (maturing). The increased number of dividing cells resulted in a dramatic increase in the size of the cerebral cortex. Different modifications of the β-catenin could almost certainly be done that would reduce the effect on the number of precursor cells that continued to divide. In addition, there are many other genes and proteins involved in the regulation of brain development that might be perturbed to increase brain size and/ or cortex size, and possibly dramatic increases in intelligence. In other words, increasing the size of the Human brain (and presumably increasing processing) appears technically very realistic.

Chenn and Walsh's study also illustrates the interesting findings of gyri and sulci formation occurring "spontaneously" in mice by simply increasing the number of cells in the cerebral cortex. In other words, the wrinkled cerebral cortex noted in Humans and not found in mice may result spontaneously as a means of packing more area of cortex (more cells) into a limited space. It has always been an interesting observation that the pattern of the gyri and sulci of the Human cerebral cortex are all different. Although the overall plan of lobes such as the frontal, occipital, temporal and parietal are the same, the "wrinkling" of each brain is slightly different.

In the movie, *Field of Dreams*, he (Costner), built it—a baseball diamond in the middle of a cornfield, and they came—baseball players began stepping out of the cornfield onto the baseball diamond. It was a dramatic and happy ending—they came!

Unfortunately, at the present time, it is not nearly as clear that if we "build it"— a larger cerebral cortex—that there would be a happy ending and a dramatic increase in intelligence. Maybe "they" would not come. As noted above, individuals with microcephaly have mild-to-moderate mental retardation, however, it is not clear at this point that simply enlarging the cortex will necessarily increase intelligence. In addition, there is currently minimal enthusiasm for doing experiments in Humans to increase intelligence, despite the fact (in at least the author's opinion) that it is clearly needed. An increase in intelligence is one of the factors that will be a major component in the future evolutionary development of mankind.

We should also note that in Chenn and Walsh's experiment all the animals died. They did not give any specifics on the cause of death; however, I am not surprised. If you suddenly have a brain that is 4 or 5 times as large, it must be adequately perfused (have an adequate blood supply). However, the heart has not gotten bigger, and it may not be able to adequately perfuse the much larger brain. If the brain is not adequately perfused, it will undergo necrosis (die). That is what is called a stroke. If the heart tries to perfuse this larger brain, it may fail. That is already a common cause of death in Humans called congestive heart failure. Therefore, increasing the brain size would require increasing the size and capability of the heart. Then there is the support problem. How could a Human (or any animal) support a head that is 5 times larger? It would require larger neck muscles and larger sturdier vertebra.

We could go on-and-on, but the important point is that you cannot simply make some major change in size to an organ like the brain without considering the whole animal, and what else needs to be enlarged or modified to support those changes.

However, despite the limitations and other factors that will have to be considered, an increase in processing capabilities, particularly by increasing the size of the cerebral cortex (scaling), will be a possible likely way of increasing Human intelligence (however, a considerable amount of additional research would be needed to even begin to accomplish this). Although increasing Human brain size would appear to be very "doable," (perturbing the concept of the cortex might [theoretically] be done by adding another layer of cortex [six cell layers] that synapses with the current cortex or increasing the number of layers in the current cortex) all of this is much more speculative. However, as we have noted, this is exactly what "Mother Nature" has done in the past.

Another interesting question is how would it be possible to organize a larger cerebral cortex? Interestingly, much of the organization in terms of synapses that occur in a newborn baby result from learning. Initially everything is connected to everything else that it can connect with, and "learning" is a process of cutting connections. In fact, learning may be at the bedrock bottom of intelligence. In other words, the brain possesses at least some capability of organizing itself, probably based on learning and use. Shortly after birth there is a widespread pruning of cerebral synapses with death (decrease) in the number of newborn neurons. This self "pruning" would raise questions about any simplistic attempt to increase intelligence simply by randomly increasing neuron cell number. Any attempt to increase intelligence based on increasing the volume of any particular area (cortical or otherwise) would have to be well-studied, and perhaps require a better understanding of intelligence in general.

Memory

The average Human short-term or working memory extends to only seven numbers (average, with a few talented individuals capable of holding 8 or 9 numbers or "things in working memory at any given time"). Although seven items in short-term memory may be enough to set an Olympic short-term memory record among living creatures, it is rather pitiful in terms of the possibility. Seven numbers—so much for mankind's Olympic record in short-term memory; perhaps we should consider another Olympic event. It really makes one wonder how we have come so far with so little. It also points out how far we must go, and what incredible advancements might lay ahead with even small incremental increases in intelligence. I am clearly using working memory here as something of a "fall guy" for intelligence mainly because it is easy to see both the current limitations and the feasible scalable increases that might be possible in working memory. Working memory is, of course, only one component of intelligence, however, it is a vital component. Memory is a component of learning, and the basis for maintaining any knowledge base.

Memory would appear to be the most scalable of all of the components of intelligence. There is nothing in principle that should prevent a future increase in memory capability—scaling. This is particularly relevant for any advances in intelligence because memory is central to all the other intelligence functions that we are discussing, such as learning or a knowledge base.

The greatest current known intelligence can (on average) only remember seven numbers in short-term memory—a U.S. telephone number. That is truly a feeble performance. Perhaps it should be written as the epitaph of mankind; "the creature whose short-term memory was seven numbers." It speaks volumes. No additional comments are necessary—the limitations of man's intelligence are clear. A future advanced or alien intelligence would understand immediately where man resides on an intelligence scale (to be discussed further below).

However, there are remarkable exceptions to the "average" that illustrate what is possible. There are currently 36 Grand Masters of Memory. To attain the rank of a grand master of memory, it is necessary to memorize 1,000 digits in under an hour, the precise order of ten shuffled decks of cards also in one hour, and one shuffled deck of cards in less than two minutes. Like an IQ of 190, these Grand Masters of Memory show what is possible even with the current Human brain. It also shows that memory can be "exercised," and potentially increased by training. Does that mean that intelligence can be increased by training? This has not really been demonstrated, but that may be a limitation on how we are currently testing intelligence. We do not seem to be giving due credit to memory as an important function of intelligence. Memory did not even make Gardner's list of multiple intelligences. Yet, I would opine that memory is central to intelligence, and should be an important component of intelligence tests, and possibly even at the top or Gardner's list.

There is also the example of Kim Peek. Peek is a savant with an incredible memory. He has memorized 9,000 books…note that is "memorized," not just read. In the area of memory, he is a genius. However, his IQ score is only 87, and he has a limited capacity for abstract or conceptual thinking. He is an excellent example of how important memory is to intelligence, and what is capable even with the current Human brain. He also illustrates the fact that memory is only one component of intelligence (we will address this shortly). What is even more incredible is that Peek was born with an encephalocele (a congenital anomaly consisting of protrusion of brain tissue through a hole in the skull) which resolved spontaneously but left a significant portion of his central brain tissue missing. Peek may have the best memory of any Human…ever. How can someone missing parts of a Human brain have such an incredible memory? The simple and correct answer is, we do not know. However, it makes me wonder if there is not a brain gene equivalent to the Myostatin gene. As we previously discussed, there is a gene that codes for a protein that suppresses muscle growth. This gene is widespread from mice to men, and the evolutionary "rationale" for such a gene appears to be to limit a creature to only as much muscle mass as it needs to make a living and survive. If you have more muscle, you have to "feed" that extra muscle. Remembering 9,000 books verbatim is ultimately going to cost energy…you have to "feed" it. The Human brain already uses 25% of our energy requirements to sustain our large brains. Thinking and memory are metabolically costly. Perhaps parts of Peek's missing brain include a part that limits memory…to make the brain more efficient. To my knowledge, no "memory suppressor" region in the brain has ever been described. However, forgetting may be more of an "active process" or perhaps the process is simply like muscle mass…use it or lose it. Some of the Grand Masters of Memory are reported to train for the event just like athletes' train for an event.

However, despite these examples of incredible feats of memory with the current Human brain (or even a damaged brain in the case of Peek), the average Human IQ remains 100, and the average short-term memory remains seven numbers or items.

What about long-term memory? Most of us have a few memories of our childhood, which might even be 50 or even more years ago—not bad. However, before you get all pumped up about Homo sapiens' memory ability, let us ask a question. How would you like to take a final examination in a college course that you took that required a lot of detailed memory a few years out of college? In college, I took "the nightmare course" of the pre-medical students—organic chemistry, about 40 years ago. That course was like trying to memorize a telephone book with endless formulas and molecular structures. In fact, many pre-medical students thought that the only "real" purpose for organic chemistry was to flunk out pre-meds. The professor would enter class and begin to write on the left-hand blackboard. There were three long blackboards across the front of the auditorium. When he arrived at the end of the blackboard on the right, he would again start on the left erasing the previous lines of organic chemical equations and writing new formulae. He used a microphone, because it was a large class, and I am not sure that I remember his face, because he always had his back to the class writing organic structures on the blackboard. That was organic chemistry—day after day.

By the end of the class, I had three notebooks full of organic chemistry formulae and structures, which I memorized exactly as in my notebooks (relatively small print and both sides of the page), formula-by-formula, page-by-page. Before the final examination, I wrote out all three notebooks line-by-line and page-by-page in three new notebooks without referring to the originals. I set the curve in organic chemistry that semester. Are you impressed? Don't be. Perhaps, I should also tell you that at the present time I can barely remember what a benzene ring looks like, and I can still write the formula for methane (CH_4) (which is probably on the first page of most organic chemistry books), and that is about all I remember. Oh yeah…there were aldehydes (HC=O), carboxyl (O=CH-OH), and alcohols (H_2C-OH), and I remember the auditorium and the blackboards—so much for Human long-term memory (or at least my long-term memory). What we know well at one time…fades fast. We may recall the incident, but the details are gone. However, if I found those notebooks, the information would undoubtedly still be there. Better long-term memory retention would be almost an absolute necessity for any advances in intelligence. Clearly, we have methods for retaining information—data (the notebooks), but they are limited, fragile (I have lost my organic chemistry notebooks) and retrieval is slow.

More recently tremendous amounts of data are being stored on computers, and it may be argued that in a general sense such data maintenance and retrieval is contributing to an increase in man's intelligence. The main problem is in the interfacing (what I call neural interfacing, which we will be discussing in the next chapter) or how we interact with masses of data stored in computers.

However, memory appears to be a critical component of intelligence that is probably scalable (it is probably also "learnable," and when we learn more about exactly how we store and recall memories (data), we may even find that there is a Kim Peek suppressor nucleus (which he does not have), which may be perturbed to give incredible increases in memory even with our current cerebral cortex and brain.

The hippocampus is a part of the limbic system, and it is intimately involved in both short-term memory and laying down long-term memory. One possible route to scaling memory might be to increase the size of the hippocampus. However, learning is highly dependent on memory, and is also largely anatomically located in the hippocampus and limbic system. In fact, even emotions are anatomically located in the limbic system even though I have given them separate status as components of intelligence. In addition, I have grouped emotions with social intelligence for some functional reasons even though emotions are anatomically more associated with the limbic system, and social intelligence maps anatomically more with the orbitofrontal cortex. This illustrates the difficulty of separating components of intelligence by function, which may not map appropriately to the anatomy. However, we will defend this grouping shortly.

How we mix the stew is not entirely clear at this time. However, I would opine that however it is done; any dramatic increase in intelligence will also require a dramatic increase in memory.

Problem Solving, Deductive Reasoning, Mathematics, Logic and Abstract Symbolic Manipulation

Mathematics, logic, deductive reasoning and abstract symbol manipulation are intimately involved with intelligence. In fact, all require the abstract manipulation of symbols which has often been equated to intelligence per se. In other words, from some perspectives they are the essence of intelligence.

When a mathematician "discovers" a new relationship or theorem in mathematics, is this a discovery of something that has always been there, but simply not known by Humans, or is it a creation of a new entity that never existed before? For example, the **Pythagorean theorem: "In any right triangle, the area of the square whose side is the hypotenuse (the side of a right triangle opposite the right angle) is equal to the sum of the areas of the squares whose sides are the two legs (i.e. the two sides other than the hypotenuse)."** (http://en.wikipedia.org/wiki/Pythagorean_theorem)

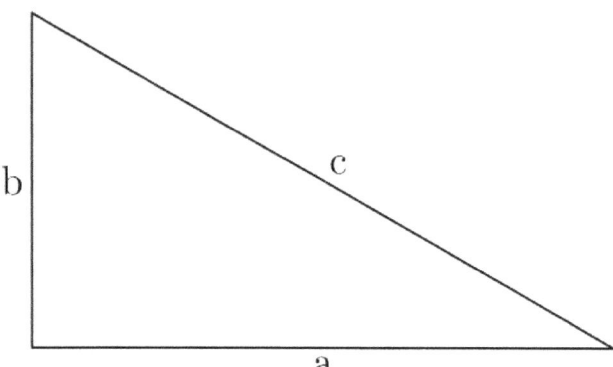

The Pythagorean Theorem: The sum of the two squares of the lengths of the legs (a and b) equals the length of the square on the hypotenuse (c).

Image Credit: Wikimedia, Public Domain
(https://commons.wikimedia.org/wiki/File:Proof-Pythagorean-Theorem.svg)

We generally learn this in school as $c^2 = a^2 + b^2$ or the length of the hypotenuse of a right triangle is equal to the square root of the sum of the squares of the remaining two sides.

Was the Pythagorean Theorem invented or was it there all along, and only discovered by man. Mathematicians have come down on both sides of this question. Leopold Kronecker, a 19th century German mathematician, noted that "God made the integers" and "All else is the work of man." (http://digitalphysics.org/Publications/Cal79/html/cmath.htm) In other words, mathematics is invented. On the other hand, the Pythagorean Theorem (relationship) would have been there even if mankind never became smart enough to discover it…therefore, it existed and was discovered. In other words, the Pythagorean Theorem "existed" when the dinosaurs roamed the Earth!

The argument is partially semantic revolving around the precise definitions of "discover" and "invent." However, I would prefer to have my cake and eat it too, and I would opine…both. There is no question that the Pythagorean theorem (relationship) is an intrinsic part of a right triangle, and therefore it exists, whether mankind knows of its existence of not. However, it is also clear that it must be discovered by a reasoning mind. In other words, without intelligence (from a thinking and reasoning mind), it would only have a metaphysical existence, and it is the intelligent mind that must invent (discover, conceive) it. We might say that mathematics is an integral part of intelligence, and likewise, intelligence is an integral part of mathematics. However, when someone is doing a mathematical calculation; what is going on? We are manipulating numbers or abstract symbols that we have learned. Is this manipulation of abstract symbols simply learning and memory mixed in the stew or is this manipulation a separate and distinct component that we name as intelligence.

I picked the Pythagorean Theorem because it is very ancient and was used by Socrates in teaching/learning to give insight into the concepts of discovery and existence. Pythagoras lived from approximately 569–475 B.C.E. (Wikipedia). Although Pythagoras is generally credited with the theorem, it was probably known in other cultures earlier. Proofs of the Pythagorean Theorem have become a cottage industry among mathematicians and wannabes of all stripes. There have been reported 520 proofs of the Pythagorean Theorem (*The Guinness Book of World Records*). It is proposition number 47 in book one of Euclid's *Elements* (written around 300 B.C.E.).

In Plato's dialogues, *Memo* (written 385 B.C.E.), Socrates is teaching a young Greek youth (Memo). The youth is impatient...difficult; he wants Socrates to teach him virtue. Socrates says: "Let me tell you an old legend believed by religious sages." "They say souls are immortal, and thus have seen and learned everything under the Sun. Deep within us, we already know everything, though during our earthly sojourn we've forgotten just about all. But if we are energetic enough, we can overcome this ignorance by recollecting it." (Crease, Robert P. *The Great Equations*, W. Norton & Co., 2008.) Socrates tells Memo that he will demonstrate the process. Socrates goes on to teach a young slave boy to reason out the question of doubling the area of a square, which involves the Pythagorean Theorem (but he does not know that). This dialogue is often used as an example of both excellent style of teaching and one's ability to reason out solutions (and perhaps to learn some virtues). The slave youth suggested doubling the sides of the square as a solution. Socrates does this, but it is obvious that the area of the square has been quadrupled not doubled. However, Socrates introduces a diagonal, which cuts the area of the original square in half. If the diagonal is extended to the quadrupled area, it cut that area in half...the area of the original square has been doubled!

However, imbedded in the problem-solving, deductive reasoning, mathematics, logic and abstract symbolic manipulation is a great deal of processing, memory, input, learning and use of a knowledge base. Even though this component of intelligence has generally been considered the essence of intelligence, it leans heavily on many other components. For example, learning simple arithmetic requires a learning process plus memory. Abstract symbolic manipulation also has a lot in common with language, which is a form of abstract symbol manipulation.

However, ultimately, we see intelligence as an emergent phenomenon arising out of the mixture of all of these components, but perhaps we need a Socrates to draw this out of us.

Knowledge Base

Most authorities on intelligence do not consider the knowledge base (education) of an individual as a part of intelligence. Instead, they consider the areas that we have just addressed such as manipulation of abstract symbols as the essence of intelligence. I must strongly disagree. Our "database" is what we must manipulate, and it is the sum of what we know and have learned. It has been estimated that there are about 10^{18} bits of information in our libraries. How many of the 10^{18} bits of information in our libraries does any one of us know? These 10^{18} bits of information are most of the cumulative written knowledge of Humankind. It is the cumulative wisdom of the wisest of the wise monkeys. Although we might make a case for Humans as having 10^{14} connections (one million billion) and equate that to computing power and memory of computers (as Ray Kurzweil has done in *The Age of Spiritual Machines*, Viking Penguin, 1999, as noted above), the question is can one synapse really store one bit of information?

That type of scenario works fine for computers, but biological storage of information appears to be much different. The Human brain has some similarities to a computer, but the differences are even more dramatic. Therefore, we need to be a bit skeptical about some of these types of comparisons. We Humans think differently, despite the fact that we are both using logic. Also, we must be cognoscente of the fact that a great deal of our nervous system is devoted to rather mundane (but important) biological functions like managing our body. Things like muscle movement, breathing, heartbeat, and other very important functions like remembering where we are, and where we live. In addition, most neurologic functions involve many neurons doing essentially the same thing—in other words there is a lot of redundancy. Therefore, most of our brain is not available to store the knowledge base of Humankind. This is just another way of saying that we are a biological entity, and not a computer. We mix the two at our own peril.

We should also recognize that our brain and intelligence did not evolve to do abstract mathematics and to learn the database of Human knowledge. Intelligence evolved to enhance survival—which means find food, shelter, avoid being killed and to reproduce. Intelligences that arose in the past that did not fulfill those basic functions are no longer with us.

The short answer to the question of what part of Human knowledge anyone knows is very little. Ray Kurzweil has estimated (*The Age of Spiritual Machines*, Viking Penguin, 1999) that the typical Human may know 100 million **(10^8)** "chunks" of information which Kurzweil loosely defined as "bits of understanding, concepts, patterns, specific skills." (My gut reaction is that this is an overestimate, but let's use his estimate). Let's guess than the most knowledgeable person in the world knows 10 times as much information as the "typical Human." That implies that the most knowledgeable person in the world knows only a small fraction (109 (109 0.00000000001% (1 x 10^{-11}%) of the current available database, and, in addition, we are probably forgetting parts of that knowledge database as fast or faster than new parts are learned. In addition, knowledge of the world is increasing at an incredible rate. That places mankind in an incredibly losing position in terms of the knowledge known compared to knowledge available. Not only do we know very little of the available knowledge, but we are also becoming increasingly dumber (when defined as the bits of knowledge "available," compared to the bits "known" by any individual. In addition, we are becoming even less knowledgeable as time passes because the rate of knowledge production is increasing.

But isn't that exactly what we have libraries for—to hold our database of knowledge. Yes, but that is also exactly where the problem of **interfacing** comes into play. Books and libraries are our current methodology for compensating for our poor memory, and the technique is rapidly reaching its limit. We can continue to store more information but finding the information and using the information in a timely manner (interfacing with the information/data) is reaching its limit. We currently **interface** with a library (the database) by going to the library, then finding and reading the information we are seeking. This is slow, tedious, and inefficient. This approach is clearly not the ultimate process and methodology for the most intelligent creature in the Universe. I would suggest that such a method of interfacing is one of the current severe limitations on Human intelligence. Information may be available, but the method of assessing it is incredibly slow.

What about computers? Information is rapidly being digitized and stored in computers. Will computers eliminate some of this problem? In the short-term, the answer is … yes. However, in the longer term, if we are to dramatically increase mankind's intelligence, it will probably require a much more integrated approach between the Human mind and computer databases (we will be discussing this more in the next chapter). The problem is not with computers or with computer memory or "recall," but again with **interfacing**. Interfacing is the rate-limiting (and intelligence-limiting) factor in the process. We currently interface with computers using a cathode ray tube or more recently, a plasma screen, a mouse, a keyboard, sometimes voice commands, multimedia-sound, pictures, and video. However, we do not have a rapid, reliable, and direct method for interfacing between an inorganic system (computer), and an organic biological system (mankind) that includes processors, memories, input sensors and output mediators. The type of interfacing that will ultimately be needed is bionic, and it's possible that the major increases that we are envisioning (giving at least the higher ends of IQ like $^{b}I_{ab}$ **1,000,000 (Homo Deus,** which we will discuss shortly)) will not be possible until organic-inorganic interfacing (which we will be calling "neural-interfacing") is possible. We will need huge storage facilities of a computer to store all of the known information of the world, but in addition, we will also have to think about the computer-- not separately. In other words, we cannot just use a computer to store information; we will have to interact with it in such a way as to think as one. You might say that we are going to have to learn how to think **within the box** (we will be discussing this more in the next chapter, *The Dumb Machine*).

Why can we be so certain of the inevitability of increased intelligence and the evolution of Homo sapiens into Homo intelligenticus that probably will begin in the relatively near future? One argument in favor of such a scenario is the history of evolution. Intelligence has literally bred more intelligence, since the "invention" of the nerve cell. We are, of course, using the generalized concept of intelligence in which a cell receives a stimulus from the environment and passes a signal to another cell, which then produces a response. This is (in its most basic form) what a neuroscientist would call a reflex arc. Through evolution, the process has been replicated, refined, and enormously expanded to produce what we now call intelligence. This is done by placing more and more neurons (processing units) between the input neuron and the output neuron.

In fact, the process of increasing Human intelligence may be going on right now. James R. Flynn has studied IQ scores of many different populations around the world, and has noted an increase, on average, of 15 points (one standard deviation) per generation. We are already slightly more intelligent, as noted by the Flynn effect, although this is a minor effect compared to what will probably happen in the future, and the genes governing specific areas of brain development, neuron number and organization (especially the prefrontal cerebral cortex) as we have reviewed are being delineated now. Computer intelligence is growing almost astronomically and techniques for interfacing between Humans and artificial intelligence are a current area of intense research which will only increase in the future. Even the incredibly slow and poor techniques of interfacing with computers and their extensive memories and rapid processing speed that we are currently using such as a keyboard, CRT and mouse have probably increased Human intellectual potential at least as much as the Flynn effect already. Consider what effect on the absolute (I_{ab} to be discussed shortly) scale for Human intelligence a rapid and "intimate" interfacing technique between biologic systems and computers such as IBM's Blue Gene with one petaflop (one million billion operations per second) of computing power will have. In fact, the potential is almost limitless. And the memory that could permanently store everything in the Library of Congress is here now. Such a technique will almost overnight increase Human intelligence by a much greater factor than the change from Australopithecus to Homo sapiens. Genetic engineering also has considerable (although perhaps not quite as spectacular) potential. We discussed mouse genetic engineering experiments that increase the size of the mouse cerebral cortex and memory (and perhaps intelligence) by a factor of 4–5.

Will it be genetic engineering or neural interfacing, which will be the more important mechanism driving **IDE**$_H$ and markedly expanding our database (knowledge), that will ultimately create Homo sapiens intelligenticus? The answer is probably both. They are not mutually exclusive, and there is no reason to expect that both techniques will make an impact on the future of intelligence. Predicting what is already happening does not take a carnival fortune teller with a crystal ball.

Learning

A newborn baby is born with a complete Human brain with almost the adult number of neurons (estimated between 86–100 billion) with approximately 100 trillion synapses. (Ackerman, S. *Discovering the Brain*, National Academies Press, 1992; https://www.ncbi.nlm.nih.gov/books/NBK234146/) This can be compared to the elephant with an estimated 251 billion neurons or the chimpanzee with 22 billion neurons. However, the elephant is not ten times smarter than the chimpanzee. This clearly shows that even though the number of neurons for intelligence is important, it is not the only single underlying factor. In other words, the number of neurons is important, but there are clearly other important factors involved in intelligence. (Herculano-Houzel, Susan., Marino, L. *Brain Behav. Evol.*, vol. 51, pp. 230–238, 1998.)

However, a newborn baby's brain is not completely "organized." What is the intelligence of a newborn baby? That is a rhetorical question with the clear answer that a newborn baby has a few "reflexes" such as sucking, and a few "built in" behaviors such as crying (we used to call them "instincts"), but otherwise babies show very few signs of intelligence. In other words, a newborn baby is not born with the capability to speak or understand a language or do mathematics or solve problems. Despite having the neurons and the basic brain structure, all areas are either not functioning or intelligence has to be learned. Babies do not remember things before about the age of three, because the hippocampus that is important in laying down long-term memory is not fully functioning. Likewise, the language areas do not become active until about eighteen months, and Wernicke's area (used for word understanding) and Broca's area (used for articulation) matures, leaving a young child in a period when they can understand what is being said to them, but are unable to respond. (Carter, R. *Mapping the Mind*, Berkeley, CA: University of California Press.) However, once the various areas of a young child's brain come "online," the important process of increasing intelligence which involves learning begins.

All functions that could be equated to intelligence have to be learned. When does a Human baby develop intelligence? It becomes intelligent only after years of learning and training, beginning with small incremental steps such as learning to control the muscles of the arms and legs, and the larynx and tongue to make sounds.

This suggests that intelligence is something that we learn, and the one "intelligent function" that a newborn Human baby is born with is the ability to learn and remember what it has learned. However, there is an important distinction here. Is intelligence in the **ability to learn** or is it in **what has been learned**?

Learning (and remembering) is an important key intimately related to intelligence. However, learning is one component of intelligence that is clearly a composite and is highly dependent on other components such as memory, and processing.

Learning would be impossible without memory, and at least some input of what is to be learned data, a complex motor skill or whatever. In addition, information, or skills to be learned would generally need to be processed. However, as noted, all these other processes are also components of intelligence. Therefore, although we will consider learning a fundamental component of intelligence, it is composed of a number of other components that are perhaps even more fundamental. However, at the present time it is not possible to characterize learning simply by expressing learning in terms of simpler or more fundamental components without losing at least a portion of the "learning" component itself. Again, learning and memory blend into one another, and it becomes hopeless to separate one from the other. If you cannot remember what it is that you learned, then you have not "learned" it. Neither learning nor memory are intelligence per se, but both are important components of intelligence, and learning is at least partially composed of memory, input of what is being learned, and some processing. In other words, intelligence is complex.

There is also the important component of understanding. Although some might consider this the essential component, I am treating understanding in intelligence as an emergent phenomenon that arises out of the components we are dissecting out.

(http://www.web-books.com/MoBio/Free/Learning/Memory.htm)

How could we increase learning? That is probably a key question to increasing intelligence. It is well recognized that children have an incredible capacity to learn language. In fact, there appears to be a "window of opportunity" to learn languages. Generally, a native speaker does much better with a language than someone who tried to learn the language later in life. Are we "dropping the ball" by not teaching our children three or four languages while they are young?

Unfortunately, as every parent knows, children also have a very short attention span, and getting them to cooperate in learning is an art. Teaching a language must be done within the practical setting of living. However, learning appears to be central to the "nurture" part of intelligence.

Language has been considered by some to be at the basis of thinking. Steven Pinker from MIT has suggested that language is an instinct, which he compares to a spider building a web. No one "teaches" the spider to make a web, and they can build an appropriate web without ever even having seen a web before. There are over 6,000 Human languages, and clearly no specific language per se is instinctual. In fact, no child is born speaking English or Spanish or German or Russian or Chinese. However, every child (neurologically normal) will learn to understand and speak a language. In fact, at about age three, children are language learning machines. It is not any language or word that is instinctual, but the process of assigning meaning to sounds (words), and then arranging those words into meaningful sentences is instinctual. However, this "talent" for language is not endless as is obvious when an adult tries to learn a new language. Many adults in learning a new language will continue to have an accent and never get the pronunciation of some words correct or fumble with the grammar and syntax.

In other words, there is a "window of opportunity" for learning languages that is evident in three-year-olds but is gone by the twenties.

Steven Pinker notes: "Language is a complex, specialized skill, which develops in the child spontaneously, without conscious effort or formal instruction, is developed without awareness of its underlying logic, is qualitatively the same in every individual, and is distinct from more general abilities to process information or behave intelligently." Pinker considers language: "man's most important cultural invention, the quintessential example of his capacity to use symbols, and a biologically unprecedented event irrevocably separating him from other animals." (Pinker, Steven. *The Language Instinct*, Harper Perennial edition, 1995.)

When we think of instinctual behavior, we generally consider it to be "hard-wired" into the nervous system. This is, in fact, the case with language, and shows that language abilities are probably quite old and very important biologically. There is a dedicated area on the left superior temporal gyrus called "Wernicke's area" that is associated with the ability to understand spoken language. Damage to this area (from a stroke for example) results in Wernicke's aphasia. Individuals with Wernicke's aphasia can no longer understand language, although they may retain the ability to calculate and do mathematics or other intelligence functions, and they can generally speak fluently. However, what they say is generally meaningless, showing the importance of Wernicke's area for processing the meaning of words and sentences.

In addition, Broca's area—left inferior frontal lobe (adjacent to Wernicke's area) is involved in motor production of language. Damage to this area results in an inability to speak a language normally, but they can understand what is said.

Although there is a tendency to equate thinking with language, there are deaf adults that have never heard a word, and who lack any form of language…no sign language, no written language, no speech, no lip-reading, and they appear to develop their own language.

In fact, even deaf children using sign language appear to use Wernicke's area to "process" and give meaning to their signs just like a language.

Despite the rather impressive use of language, and its contribution to intelligence, there appears to be other ways of thinking or at least ways of compensating when language is lacking.

However, there are some very basic questions that are heatedly argued within the various fields of research but appear to remain unanswered at the present time. For example, is someone lacking a language because of a handicap, such as a hearing problem, also handicapped in terms of intelligence? The politically correct answer would be no; however, the politically correct answer may get votes, but is not necessarily the correct answer. In other words, we are not asking can they function, or do they demonstrate some intelligent function; the answer appears to be yes and may reflect the incredible plasticity and ability to compensate by the Human brain. However, would the same person (brain) have the same IQ or a much higher IQ, if they had language added? Our compassion for the handicapped almost demands that we not ask the question for fear of the answer.

However, our quest remains: "How can we increase the intelligence of mankind? Can we increase it by increasing the language function?"

Social Intelligence and Emotional Intelligence

It might seem odd (and it probably is) to consider emotions as a form of intelligence. However, emotions like fear may have been both lifesaving and one of the earliest forms of "thinking." In addition, most of our emotions map to the limbic system, which is considered by some to be the older (early) mammalian brain (cortex). Although "emotional thinking" is now under the "control" (or at least modification) of the cerebral cortex, that control is not absolute, and emotion such as anger clearly do not always correlate with logic or what might be considered intelligent thinking. The probable future of this area will be to try to gain better control of our emotional thinking. In other words, in this area, less may be more. However, it is also emotions that make Humans…well, Human. Our general mood and important Human concepts such as love, compassion and many others are emotions that are central to Humanity. The problem may be separating the chaff from the wheat. In addition, there is an "award system" that literally makes life worth living. This reward system consists of a group of dopamine neurons in the ventral tegmental area (in the midbrain) with connects to the nucleus accumbens septi in the forebrain, and both connect to the orbital frontal cortex and help us control impulses, apply cultural mores and understand the consequences of our behavior. These areas are also involved in some forms of addiction.

We do not want to increase our intelligence at the expense of losing emotions and behavior that we generally consider to be at the base of Humanity, and what also makes life worth living. Our emotions have a considerable impact on social intelligence, and the way we interact with others individually and collectively.

Theory of Mind is also a part of social intelligence. We discussed that earlier. Theory of Mind develops at about age five. Is it something learned, or does it result from some "maturation event" in the brain? I would opine that it is a "learning skill" like most forms or facets of intelligence. However, as we have already noted, learning relies on almost all of the other components of intelligence, including memory, processing, input, and an anatomic feature of the brain itself (which I have roughly correlated with hardware).

Social intelligence might, on first pass, also not appear as important to intelligence per se as other components such as memory, processing, etc., however, social intelligence appears to be another of what we have called "value-added emergent phenomena" (we have seen this many times before, and we have used the metaphor of the disassembled car parts versus a running "functional" automobile), and it is at a different level of neural organization and function. This level of organization affects behavior directly rather than functioning at a more basic level of neurological organization.

The main idea embedded in the concept of social intelligence is the social interactions among people. There is a part of the brain that is specifically used for facial recognition. Facial recognition could be considered a part of social intelligence, since it is critical that we recognize our mother or our family versus a stranger. Social intelligence gets much more complex the larger and more intricate the interactions that we have with others. In fact, the complexity of the social interactions has been suggested to have been one of the "selective pressures" that may have advanced overall Human intelligence. However, I wish to emphasize that social intelligence is functioning at an entirely different level of organization than any of the other components that we have discussed and is much closer to the concept of thinking. It affects behavior directly. In addition, it appears that at least some other animals such as chimpanzees and other great apes also display some aspects of social intelligence (in addition to other intelligent components such as memory and learning).

However, our quest here is to inquire as to how we might enhance our social intelligence in the future.

One possibility is to recognize that the "social interaction" between small groups has now expanded to a social interaction of the entire world, and these interactions continue to be more complex. We have briefly discussed this evolving social interaction in the *Manifest Destiny of Mankind* as a "higher order" of organization that is roughly called civilization. It includes such areas as democracy as a form of government, and a constitution telling a government how it must operate and what powers it has and does not have, but also recognizes the potential of democracy to be nothing but the tyranny of majority groups (or even individuals), and requires a clear understanding and the building of Human rights and Human needs to be the central basis for moral and ethical decisions and behavior. However, it is critical that not only does the world need social intelligence at its core, the generalization of social intelligence to the level of civilization will be the next "level" of organization or evolution of social intelligence. People on the other side of the world are now our neighbors, and we must all live by the Golden Rule.

We are, in general, not used to speaking of things like Human rights as involved in intelligence, but I would opine that this is a legitimate consideration. Human rights are simply the "Golden Rule" applied uniformly and equitably to all people on Earth. Although social intelligence is generally considered to involve the interaction between people, social interactions have now blossomed to include interactions between large groups—nations. An individual's morals and ethics are often imbedded in a religion and are often given to an individual during his/her childhood from our parents, family, church, and school, and seriously affect our emotions and emotional thinking, and our interaction with others. In other words, social intelligence has "evolved" and no longer just involves individuals but the interaction of large groups…all over the world. In addition, we cannot simply try to re-engineer one brain, we must re-engineer the whole society with values such as respect for democracy, Human rights including slavery, women's rights, children's rights, and a whole table of incredibly important concepts that must become the core foundation of social intelligence and universal moral and ethical behavior and social interactions.

How could we possibly accomplish such a feat?

Since social intelligence appears to be an emergent phenomenon resulting from a "higher level" of organization of many components of intelligence, including emotions, we may achieve some "increase" (whatever that means) as we scale other components of intelligence. Possibly like what we noted with consciousness. However, applying social intelligence to civilization, in general, is going to require a continued world effort. Such a continued effort will be equally as important as conquering disease, extending the lifespan of mankind or increasing mankind's intelligence, and, I regret to say…probably much harder to achieve. However, the important concept here is that the generalization of social intelligence will involve not just the interaction between a few individuals, but large groups of individuals organized often into motley groups such as nations, religions, races and every other permutation. The future of intelligence and of mankind itself requires that the development of social intelligence be a high priority. Perhaps, we can find a way to insert the "Golden Rule" into every Human. This is probably going to be the most difficult of our components of intelligence to advance, but likewise possibly the most important and rewarding.

Output

Muscular contraction is the most basic form of observable "output." Muscular contraction is controlled by nerve stimulation. The relationship between nerve stimulation and muscle contraction is so intimate that a muscle cell that loses its nerve cell innervation will undergo atrophy and eventually die. The sum of this neuron-directed muscular activity forms what we would recognize at a higher level of organization as a part of behavior. Behavior mediated through muscular activity is one of the most important intelligent outputs of a living organism. More advanced (evolved) organisms have also developed sound as an important output. Sound is also the result of nerve-controlled muscular activity. However, when sound was fashioned into a language, it became a powerful means of communication (output and input) and allowed the projection of intelligent functions (ideas, plans, needs, etc.) from one individual to another. As noted above, the "invention" of language increased Human intelligence. However, what we hear in language coming "in" is input, but our responses are "output" mediated by muscular activity in the mouth and throat or other muscular activity.

However, there is another form of "output" that never even leaves our brain. Manipulation of abstract symbols (words) that never are formed into sounds becomes at least one form of thinking. You literally talk to yourself while thinking and bypass the production of sound. The brain's output (in the form of manipulation of abstract symbols such as words) forms a part of the process of thinking. As we noted previously, we can think using signs or other mental objects, and these can be given meaning even in people who have never heard a word.

Consciousness

My main problem with consciousness is that we all have it, and therefore each of us has some "personal experience" with consciousness. It appears to exist (we experience it), even though we really do not know what it is. However, neuroscientists have never been able to find an anatomic location for consciousness. Francis Crick considered consciousness to be the most important unsolved problem in biology. Wlodzislaw Duch of Nicholas Copernicus University in Poland has noted: "The noun 'consciousness' does not refer to anything in particular, it is not a thing we can talk about, it is rather an experience or many different experiences that we label as 'consciousness." William James, the father of American psychology, expressed a similar opinion over a century ago, and noted that consciousness is not a thing but a process. (Francis Crick in *Mapping the Mind* by Rita Carter, Los Angeles, CA: University of California Press, 1998.)

The first difficulty in the research on consciousness is to clearly define what the real problem is." (Wlodzislaw Duch, IV National Conference on Models of Biological Systems, Krakow Poland, 1995, http://216.239.51.104/search?q=cache:wsyvNuecixkJ:www.phys.uni.torun.pl/publications/kmk/95consc.pdf+Wlodzislaw+Duch+%22IV+National+Conference+on+Models+of+Biological+Systems%22&hl=en&ct=clnk&cd=1)

This appears to clearly delineate two major problems involved in consciousness. First, since consciousness is not a "thing," it is highly unlikely that a neuroscientist will ever announce that they have located a group of neurons that are responsible for consciousness. Second, consciousness appears to be many different things lumped together. For example, we can differentiate between someone who is conscious from someone who is unconscious or asleep. This form of consciousness is often referred to as arousal and involves the Reticular Activating System of the midbrain. Damage to the Reticular Activating System can result in coma. Some neurons in this system are involved in the wake/sleep cycle which is another facet of consciousness versus sleep. In addition, the release of neurotransmitters such as dopamine and noradrenaline from the reticular neurons affects alertness of the prefrontal lobe and brain.

However, that type of consciousness (arousal) has little to do with the self-awareness of ourselves as different from other people or objects around us, which some consider to be an important part of consciousness (the "I" or myself part of consciousness).

Attention appears to be a function of the anterior cingulate gyrus, which is located in the longitudinal fissure that separates the two halves of the brain.

Probably the easiest form of advanced intelligence that can readily be appreciated would be to scale consciousness with multiple parallel consciousnesses. The Human brain is already a master of parallel processing, but almost all our current parallel processing is done unconsciously, and we are only conscious of one stream of thought during any given interval. In the advanced intelligence we are describing, there would be multiple parallel consciousnesses all at the same time. Such intelligence might be like ten or even a hundred different people all thinking at once, but in one individual, and without conflict. We are currently capable of doing some of this now—for example when we are driving and visualizing the road and following a particular route (on our way to work) and listening to a news broadcast at the same time, however, there is still only one real consciousness functioning at any given time. The concept that we are envisioning here is a considerable expansion of consciousness capability that would be quite different. Let's use the jargon "it would be emergent." Are multiple consciousnesses realistic? The increased intelligence would "simply" be a matter of scaling consciousness. There is nothing in principle that should prevent such scaling, but there is also no blueprint for how neuroscientists and genetic engineers could approach the problem of constructing multiple consciousnesses. In addition, there would be the less than trivial problem of coordinating all of the consciousness and creating a "cooperative" effect on thinking.

Multiple consciousnesses would also allow multiple inputs and outputs to function at the same time, which would be another potential mechanism for increasing intelligence.

In other words, the potential future hominoid that we are envisioning (at least an early IDE_H evolved version) would be capable of focusing attention on multiple inputs and outputs at one time and both process and coordinate such "thinking." However, my disclaimer is that we do not understand consciousness on a basic level, and projecting an advanced consciousness is, at least at this time, impossible.

We might also view consciousness as another "value-added emergent phenomena." From this perspective of consciousness there is no area we will ever be able to point to as the location of consciousness because it results from the interaction of many different brain areas. However, this also leaves open the possibility that if multiple areas that contribute to consciousness are anatomically or functionally increased, then the results would be an increase in "consciousness" (whatever that might be in reality).

Intelligence on an Absolute Scale

Perhaps we need to expand our concept of intelligence, and to consider intelligence on a more "absolute scale" like temperature. When dealing with temperature, there is an absolute bottom limit to temperature. It is called cleverly, absolute zero or 0 Kelvin (0K) or -459.67 degrees Fahrenheit (-273.15 degrees Centigrade). There is no thermal energy—temperature, less than absolute zero. Zero Kelvin is as far down as you can go—the absolute bottom. As Humans, we live comfortably at temperatures between say 60- and 80-degrees Fahrenheit (288.7K to 299.9K on the Kelvin absolute temperature scale), and we can survive in an even wider temperature range. However, the range of temperatures that we are comfortable with or even the temperature range in which we can survive are extremely narrow compared to the possible temperatures on the absolute temperature scale.

It would be beneficial to broaden our concept of intelligence and consider intelligence on an absolute scale like temperature. An absolute intelligence scale (I_{ab}) would allow us to consider many new generalized concepts of intelligence like: what would a higher intelligence of I_{ab} 1,000 be like? The scale probably should be linear and would be set so that within the current IQ range, the IQ would be comparable to the I_{ab}. (IQ = 100, I_{ab} =100, IQ =150, I_{ab} =150, etc.).

By matching the IQ and I_{ab} scales in the range of zero to 200, much of the previous work in the area of intelligence would still have meaning and relevance.

This would also allow comparisons with current perspectives regarding intelligence. However, IQ has an upper limit, based on a mean and the current smartest Human. There theoretically should be no upper limit to I_{ab}, although that is yet to be proven.

1) There is nothing **in principle that precludes higher intelligence** than is currently recognized in Humans—with a maximum intelligence of about an IQ of 200 ($I_{ab} = 200$). But what would constitute such a higher intelligence? For example, on an absolute intelligence scale, we might ask the question as to what would an intelligence of $I_{ab} = 300$ be like? In addition, the absolute intelligence scale should **not be linked solely to Humans**. It would be linked only to intelligent functions of thinking and behavior reflecting such thinking per se. Animals (or anything that demonstrates intelligence) could be assigned an intelligence on the absolute scale (I_{ab}). Intelligence would be recognized as multifaceted and composed of integrated components (discussed above). However, this time, we cannot be entirely clear that such "higher intelligences" can even exist or what it would be like!

Clearly, there is an absolute bottom or zero for intelligence similar to absolute zero temperature. An intelligence of zero speaks for itself—there is none. If a creature's intelligence is equal to that of a stone, then they are literally at "rock bottom"—zero intelligence (pun intended). There are also levels of intelligence above zero which for Humans has been set at an average IQ of 100. The history of how the intelligence quotient (IQ) came into being as a test for intelligence was discussed above.

Setting the average Human IQ to 100 was arbitrary. The scaling factor is also arbitrary. However, what is important is the ability to compare intelligence levels—higher or lower on the IQ scale. Over the years there has been a recognition of the many deficiencies of the early IQ test, and multiple new tests have been developed to correct some of those deficiencies, such as cultural bias. However, there have persisted several important deficiencies which are inherent in IQ concepts of testing as a "measure" of intelligence, and that has also affected our general view of intelligence. One deficiency is that there has been a general neglect in discussions of higher intelligence—well above the current range of Human. Any discussion of higher intelligence inevitably means Humans with an IQ in the range of 140 to perhaps 190 to 200. This is, of course, partially legitimate in that there is no IQ much above 200, and IQ testing is restricted solely to Humans. However, if we view intelligence as an entity in and of itself, and a scale to quantitate the various levels of intelligence, we can begin to perceive the possibility of higher—even much higher—levels of intelligence, and perhaps even totally new forms of intelligence. Spectrum of intelligence on an absolute scale with a zero bottom ($I_{ab} = 0$), and extrapolate our scale and scaling factor to correlate with the current IQ scale so that the average Human intelligence is set at $I_{ab} = 100$ with a maximum current Human I_{ab} of about **200**, we can now consider higher levels of intelligence like $I_{ab} = 1,000$ or even $I_{ab} = 10,000$ (if such I_{ab} even can exist).

Clearly, our current understanding of what might constitute a higher level of intelligence is completely unknown. Human arrogance, which we have previously discussed, has been at least a contributing factor in mankind's ability to have tenaciously held some very incorrect beliefs for thousands of years. Examples are numerous, but one that we have discussed is the assumption that we Humans are at the center of the Universe. This same type of arrogance has locked us into the belief that there is no intelligence higher than the top of the Human IQ scales. However, let us consider some simple mental tasks. The average Human is limited to a short-term memory of only about seven or eight numbers. Most of us learned multiplication tables up to ten. Seven times eight is fifty-six. That may be quite impressive compared to an animal, pick your choice—dog, horse, elephant. There is no evidence that these animals understand the concept of numbers or multiplication. Therefore, mathematical ability appears to have literally "fallen out" of mankind's increased intelligence.

However, what additional "intelligence" might occur with further advancements is a total guess or completely unknown.

Let's consider an absolute scale of intelligence and consider an entity that can remember a thousand numbers and can multiply 100 numbers of times 10,000 in a fraction of a second. Of course, a computer can do these types of mathematical gymnastics and a lot more even now, but we will agree (for the moment) that they are not intelligent. A computer is simply a machine. Most experts in the area of intelligence would agree that short-term memory or multiplication or even all mathematical skills are not intelligence per se, however they are a **part** of what most would generally consider intelligence. Intelligence is obviously much more, but memory is a part of intelligence. But exactly what is intelligence remains elusive—despite the myriad "definitions." Despite the fact that we all have a sense of what intelligence is or is not, it is a remarkably difficult concept to define…exactly. We all have at least a little of it, and it is unquestionably the property that is what defines Homo sapiens as a species, but we really do not know exactly what it is. Intelligence is learning ability— problem solving, memory, verbal and mathematical abilities, and a long list of other skills and talents—yet it is none of these by themselves. Intelligence is more of a process that results in and from all of these abilities and more. We face a problem with intelligence very similar to what physicists studying the characteristics of the Universe often complain: there is only one sample of higher intelligence to examine—us (just like there is only one Universe for physics to study), and we are what we are trying to understand and study.

However, to generalize the concept, despite our difficulty in exactly defining intelligence, we would suggest that there is an absolute scale of intelligence like temperature, and that we Homo sapiens—even the most brilliant amongst us— are on the very lower part of this absolute scale. Our intelligence is only a hot day as a metaphor… let's say 100 degrees Fahrenheit (conveniently set at our average IQ).

But considering the range of possible temperatures, there is a lot of room to get a lot hotter. A blast furnace used for smelting iron has a temperature of about 3,000 degrees Fahrenheit, and the temperature in the center of the Sun is ten million degrees. That is hot. With temperature, we have no problem with defining it exactly or understanding the concepts of the relative temperature difference of a hot day, a blast furnace, and the Sun. Although we may get at least some small sense of what the temperatures of a blast furnace is like standing near one, the real meaning of such temperatures is also not truly comprehendible to us as Humans. We can manipulate the numbers or compare a blast furnace temperature to the Sun's temperature, but what is ten million degrees Fahrenheit in Human terms? Similarly, we can project higher intelligence on an absolute scale, but because we also have no experience with such a concept it becomes difficult to truly understand or imagine exactly what such intelligence would be like. What is even worse is we really do not know how to even define such intelligence. As we have noted, the current IQ measure of adult intelligence is based on the concept of a bell-shaped curve distribution of the intelligence of the Human population with a mean value. Intelligence tests arbitrarily set the average Human intelligence at 100. There is, of course, nothing special about the number 100. Clearly, the current IQ tests based on a bell-shaped curve are useless above an IQ of about 205 which is 7 standard deviations above the mean of 100 with a standard deviation of 15 (Wechsler Adult Intelligence Scale).

Animal intelligence generally lies below 40…but where? Animals generally lack complex language. Try thinking about some complex problem without using language, and the limitations become rather obvious. This is why many in the past have not assigned any intelligence to animals—instead, animals have "instincts." However, this is clearly an anthropocentric attitude that can only be justified by the very narrow definitions of intelligence.

Consider a lion in Africa that picks up your scent and tracks you while keeping itself concealed in the brush. It works its way around in front of you. Finally, it watches the direction you are heading and crawls stealthily close to your apparent path, waits patiently, and when your distance is such that escape is impossible...charges and kills you. Using our current anthropocentric view of intelligence, the lion has only demonstrated some "instincts"—no intelligence. This fact may comfort you as you are consumed for dinner, but all of the complex processes of detecting your presence by smell, following your trail, remaining hiding in the brush while stalking you, working its way around to set an ambush, estimating the appropriate time to strike, and recognizing you as something for dinner are all extremely complex processes and signify intelligence. Clearly the lion has no language ability and would do poorly on the Stanford-Binet intelligence test, but that is only because such tests are designed only to recognize Human parameters of intelligence. The current Human measure of intelligence—IQ—is irrelevant and impossible to apply to the lion. However, remember, you were dinner...not the lion, which attest to the fact that the lion's I_{ab} is not zero. What is the lion's I_{ab}? It is high enough that you were dinner!

Of course, having a scale is not enough; we need to have a way of measuring the quantities that constitute intelligence—an intelligence thermometer in our metaphorical comparison to absolute temperature. This is, again, where the components of intelligence that we have discussed begin to show some additional value. Let us consider intelligence on our absolute scale as the sum of ten of the components of intelligence that we have discussed. The ten components are:

1. Input (I^1)
2. Short and Long-Term Memory (I^2)
3. Processing (I^3)
4. Anatomy and Neural Function (I^4) (in essence using computer terminology, the hardware)
5. Learning (I^5)
6. Problem-Solving and Symbolic Manipulation (I^6)
7. Knowledge Base (I^7)
8. Language (I^8)

9. Emotions and Social Intelligence (I^9)
10. Output (I^{10})
11. In addition, there is consciousness—c (and possibly g, which stands for a general term for intelligence), which we are calling a multiplier.

Therefore,

$$I_{ab} = c(g)(\Sigma\ I^1+I^2+I^3+I^4+I^5 + I^6 +I^7+I^8 + I^9 +I^{10})$$

(The Σ symbol means "sum of.") The superscript "**h**" is for Human; however, some aspects of intelligence can probably be applied to artificial intelligence (non-Human intelligence), and the superscript "**a**,") ($^a I_{ab}$) could be applied. In addition, at least some aspects of intelligence or at least some of the components of intelligence as noted above can be found in computers. We would also note that computer intelligence "**c**" ($^c I_{ab}$) if we define it properly) would be recognized as one form of artificial intelligence "**a**." These considerations and discussions become difficult since some (many) do not recognize any real intelligence in computers. Some Humans are outraged by suggestions that computers have any intelligence at all. However, computers can outperform Humans in many of the components of intelligence noted above. The problem may lie within our definition of intelligence and understanding of the relationships between Human intelligence and the components of artificial or computer intelligence.

Let's consider an example. If we assign 10 points to each component of intelligence noted above for the average Human and we multiply by a consciousness of one (and possibly a **g** of one, if it exists), then the current Human intelligence would be 100 as currently recognized. Computers lack consciousness (**c**) and probably the **g** factor (whatever that really is) would be giving a **c(g)** of **0**, therefore computer intelligence would be zero. However, some would object to that estimate, and probably within some considerations rightly so. Computers can do incredible things, but the problem is that they appear not to really understand anything at a "conscious level," so therefore $^hI_{ab}$ (h is Humans) would be **100** and $^cI_{ab}$ (c is for computers) would be zero. If we set c (and g) at 1, then 1 x 100 gives 100, and we have the intelligence of the average Human. What we want to consider is how could we increase intelligence? This should also make it clear how it might be possible to increase intelligence by increasing individual components of intelligence.

You do not like this system! Fine, you are free to define your own system. This is simply a "first attempt" to suggest a system that could be worked with to give some approximation of future intelligence, because we are about to ask a profound question!

What would a higher intelligence "look like?" How do you describe something that has never existed? This is going to be a monumental task! Undoubtedly, there will be plenty of controversy! But it is highly probable that if mankind survives that a higher intelligence will occur either spontaneously of implemented by mankind.

What would an intelligence in the thousands... say 3,000 (I_{ab} = **3,000**) be like? What would an intelligence of one million (I_{ab} = **1,000,000**) be like? These intelligences do not exist now, but could they exist in the future? Let us consider the metaphor of the temperature of a blast furnace or the interior of the Sun compared to the temperature of a hot summer day (100°F). Perhaps this would give some numerical comparisons, but does such a consideration have any relative meaning? Clearly such intelligence would be incomprehensible in terms of current Human experience, but the question at hand is whether such increases in intelligence are even possible? In principle, marked increases in intelligence should be possible. In principle, there are no reasons that there should be an upper limit on Human intelligence (or even artificial intelligence for that matter once we have defined exactly what consciousness is and whether it can be programmed). In fact, if we break intelligence down into some of its recognized components such as working or long-term memory or learning as above, computer components of intelligence would appear to be quite scalable. At the present time we do not know how to scale them, but theoretically they should by some method be scalable. If computers have consciousness, how would we access it? The simple answer would be to turn the machine on. However, that answer is not very satisfying and leaves a lot to be desired.

For example, the number of items such as numbers that the average Human can store (remember) in working short-term memory is about seven numbers (items). Let's say a person could be able to work with 21 numbers (objects or ideas) in working memory at one time (21 is 3 times the average number Humans can generally maintain in short-term memory, which is 7).

We have just scaled intelligence by a factor of 3 by increasing memory from 10 to 30 and leaving all other intelligence components as 10 and multiplying as per our intelligence discussion above. Of intelligence noted above (memory—I^2 now =30), (of course, since the components of intelligence are not all "pure," we may have also used some processing (I^3) and problem-solving and symbolic manipulation (I^6) and perhaps other components of intelligence from above like input, output, etc., but we will maintain them as above at 10. If we changed nothing else but the memory, then the $^bI_{ab}$ would be 120 (the sum of our ten components with memory increased by a factor of 3 (I^2 now =30) times consciousness (c = 1),). In other words, $^bI^2_{ab}$ = 30 and $^bI\ (120)_{ab}$ = 120. In fact, memory would appear to be the most easily scalable of all of the components of intelligence, and perhaps one of the easiest to measure. If all the components were scalable by a factor of 3, we would reach an $^bI_{ab}$ = **300**. Of course, it is not clear that we could scale all the components of intelligence noted above. We are simply considering a "what if scenario." At the present time no one knows how to do any of this! In fact. it is not even clear that many of the factors that we have dissected out of intelligence can be scaled at all!

For example, memory might be scalable simply by adding more cortical and hippocampal neurons. However, all of the current recognized components of intelligence may not be scalable (and probably are not), and some new components of intelligence may be possible that we currently know nothing about. For example, one possible "new" intelligence component might be parallel consciousness as noted above—the ability to think and maintain multiple ideas in multiple consciousnesses all at the same time. Let us say that multiple consciousnesses were scalable, and our conjectured intelligent future hominid could function with multiple parallel consciousnesses that interact, communicate, cooperate and influence one another all at the same time. Remember, increasing consciousness using multiple interacting consciousness (**c**) would be acting as a multiplier in $^h\mathbf{I}_{ab}$. This would appear justifiable, since we would, in essence, have multiple individual intelligences that cooperate and communicate with one another.

As is very apparent, even trying to imagine advanced intelligence can be very complicated, and we cannot even be sure that any of these considerations are possible.

It is even possible that as we increase intelligent components that a point might be reached where a "value-added emergent phenomenon" might occur, producing something in intelligence that we cannot currently even conceive as possible. If we again use our "well-worn" metaphor of the automobile parts that when properly assembled results in an "emergent phenomena," such as an automobile that can be used to travel, then if we perturb the parts sufficiently, we might end up with a jet plane (emergent phenomena) that could take us from New York to Paris much faster that any automobile. However, it is also possible that no such emergent phenomena may be possible, which again reflects the difficulty in predicting or prophesying phenomena such as advanced intelligence that does not currently exist, and that has never even been conceived of before.

We have also mentioned that an absolute intelligence scale using components might be used for computers and artificial intelligence or at least for making a "ballpark" comparison with Human intelligence. If we calculate the absolute intelligence of a computer, and we are multiplying by consciousness, which is currently considered zero then $^aI_{ab} = 0$. However, we can still consider a computer's components of intelligence by considering its memory, processing capability, abstract symbolic manipulations, etc. Therefore, we have another approach to measure and expressing artificial intelligence—by using the components of intelligence: $^cI_{ab}^{co}$ (co is for components such as $I^1+I^2+I^3$...as above). Again, all of this is difficult to consider when we are trying to understand or predict future intelligence!

For example, a computer that can process 17 trillion bps has clearly surpassed Humans in processing even with the massive parallel processing of the Human brain and could have a processing component of intelligence of let's say $^aI_{ab} = 1{,}000$ (however, since computers have no consciousness, their IQ would still remain at zero). This, of course, is not a valid comparison. Humans and computers "think" entirely differently. We can only make a "ballpark" comparison, and note not just the similarities, but the large differences. We have introduced the components of intelligence to be able to make a ballpark comparison in our effort to consider higher forms of intelligence.

Clearly, some of our components would have to be redefined for artificial intelligence such as anatomy and function or even emotional and social intelligence. In addition, although it is "fun" to make comparisons, we must appreciate the considerable limits of such a comparison. Yet, we are trying to gain some insight into what a higher intelligence would be like.

Using our $^bI_{ab}$ is pure speculation—yes, without question...now! However, again, in principle, there is nothing that absolutely precludes an intelligence with marked increase in working memory, long-term memory, increased ability to learn or even parallel consciousnesses or scaled number of neurons in the cortex or other areas of the brain (which would also increase our anatomic components that contribute to intelligence).

As a part of our journey in the future, and particularly as a part of our themes to view man as a **work in progress**, we are looking at future life and our Universe with a broader perspective, and we should begin to realize and appreciate that the intelligence that we are rightly proud of has not been around that long—about 150,000 years or perhaps 200,000 years (at the most)—compared to an Earth of 4.6 billion years and a Universe of about 13.7 billion years. The history of what we call Human intelligence (or any intelligence for that matter) is rather brief. What may be an even greater blow to our Human ego is the appreciation that we have an incredibly low intelligence on an absolute scale of what is conceivably possible. This would also be consistent with the conjecture that as sentient beings we are perhaps the lowest form of God conceivable or perhaps possible intelligence. The good-news bad-news aspect of that conjecture is that we are, at the best estimate, at the bottom of the God heap, and there is a lot of potential for improvement. Remember, what constitutes a "God" is all in the definition. Also remember, that if we add "faith" to the brew, then reason and even any reasonable definition of a God goes down the drain!

Increasing Intelligence by Adding a "New Cortex" (Not Just More Cortex)

One approach to imagining a much higher order of intelligence would be (as we have already discussed) to simply extend our current abilities—like memory, however, that would probably be completely inadequate to increase our intelligence into an $^hI_{ab}$ approaching 3,000. Alternatively, it is also highly unlikely that we could increase mankind's intelligence in any significant manner without accompanying marked increases in memory and processing. In addition (as we have also discussed), there is at least some possibility that if intelligence were markedly increased that an emergent phenomenon would come out of the results like life coming out of chemistry or even consciousness and intelligence coming out of a group of neurons connected together. We have seen evolution's development of brains from the primitive reptilian brain to the mammalian brain by the addition of the limbic system on top of the older reptilian brain, and the development of an additional layer of nerve cells on top of the limbic system in the form of the cerebral cortex. However, the concept of the reptilian brain with the mammalian brain added on top has been (probably rightly) severely criticized. Although not as simplistic as generally expressed in the concept of the reptilian/mammalian brain concept, it is still valid that the brain and intelligence has been evolving. Each of these developments seems to have resulted in the development of a "value-added emergent phenomenon" of greater intelligence. What would happen if we engineered an additional 6-celled layer of cortex on top of the current Human cortex? The brain has shown itself to be very adept at effectively utilizing the nervous tissue that it has available. Could a new "emergent phenomena" result from adding a newer cortex to control not household tasks such as breathing or heart rate or even limbic tasks such as memory, emotions, and learning, but interact with prefrontal cortex to expand executive brain functions. Adding something (like neurons and nuclei (clusters of neurons) "on top" appears to be how "Mother Nature" has expanded intelligence in the past. Adding a new "executive cortex that interacted with existing "executive function cortex" might act as a new multiplier of intelligence or at least a "new cortex" might increase the anatomic/functional component of intelligence. Since this approach appears to be "Mother Nature's" approach, we should not dismiss such a possibility out-of-hand. In addition, such a "new cortex," should not be involved with housekeeping or somatosensory or motor function but should be involved and interacting with executive functions and abstract manipulation of objects and language. Is such a "new executive cortex" possible or is it science fiction? We would have to leave this as an interesting possibility, but a lot more speculative than almost any other possibility except perhaps multiple consciousness. To be frank, it is more in the science fiction genre, but also recall that space travel was once science fiction.

New Forms of Intelligence

Another possible "new form of intelligence" might involve dimensional thinking. We currently think in three dimensions. The ability to think dimensionally is already recognized as an intelligence function and is evaluated on some IQ tests. However, the type of dimensional thinking we are referring to here would be multidimensional. Some forms of string theory suggest that the Universe actually exists in 10 or 11 dimensions (plus time, which in Einstein's theory of relativity is also treated as an additional "dimension" on par with the three space dimensions [space-time]). Would it be possible to think in ten dimensions?

Then there are numerous other forms of mental capabilities claimed for years by various people that appear to have no real validity but might be used here as an example of the possibility of science fiction predicting the future or at least some variant. Telepathy (sometimes redundantly called mental telepathy) is the "communication from one mind to another without the use of sensory perception (WordNet 2.0 Dictionary)." Telepathy would be a sort of "mind reading." It could be viewed as an advancement of intelligence since it would be the ultimate "Theory of Mind." In essence, with telepathy one would know exactly what someone else is thinking rather than the current version of Theory of Mind where you deduce someone else's thoughts by the situation, their behavior, past experience, facial expressions, body posture, etc. There are other types of mental phenomena such as telekinesis (the ability to move objects by means of thought alone, without physical means (Webster's Dictionary 1913), and clairvoyance (the power to perceive things that are not presented to the senses) (WordNet 2.0 Dictionary). However, scientists have repeatedly looked at these types of "phenomena," and they have concluded that they do not exist, and it is not clear how they could ever be created.

Despite serious scientific research in this area showing no evidence to support such a phenomenon, a Gallup Poll has noted that about 36% of the population still believes in telepathy. The belief in telepathy as a mental phenomenon already present in some people's brains (at least at this point in time) is an example of impaired intelligence. One aspect of intelligence that we have discussed is the ability to solve problems, use logic, and to be able to find and review data and draw realistic conclusions. In fact, it would be very legitimate to put questions regarding such phenomena as telepathy, telekinesis, and clairvoyance (power to perceive things that are not presented to the senses (WordNet 2.0 Dictionary) and many other similar phenomena on IQ tests. People with such beliefs clearly have difficulty with problem solving and reality testing, which is an important component of intelligence. People with such beliefs are clearly not that bright regardless of their vocabulary, mathematical abilities, or other talents. The ability to tell reality from fantasy is a very important basic intelligence function and should be better recognized as such even though that will not be a popular position. However, the question is "could such capabilities be developed in the future—as a future expansion of intelligence?" If we insist on the part of the definition of telepathy 'without the use of sensory perception," the answer is probably…no.

The exact path to increased intelligence is not known at this time. In fact, at the present time, we cannot be entirely certain that intelligence can be increased beyond the current highest examples that we are aware of—about $^hI_{ab}$ = 200. However, it should clearly be possible to raise the majority of the Human populations $^hI_{ab}$ to at least the current upper level of $^hI_{ab}$ = 200 which has supposedly been achieved (however, again this is based on a particular estimate of IQ that not everyone may agree on). This would be a considerable improvement, but nowhere near the improvement needed to begin to reach what I am calling **Homo intelligenticus** ($^hI_{ab}$ = Hi) with a minimum $^hI_{ab}$ = 2,000, and it fades into nothingness when we consider a very advanced intelligence like what we are calling **Homo Deus** with an $^hI_{ab}$(HD) of at least $^hI_{ab}$(HD) = 100,000. However, note the caveat that each of these $^hI_{ab}$'s are arbitrary and perhaps not even achievable or meaningful, and you are free to define them differently. We have used values that should make it possible to define a new species by changes of intelligence only (which, as we have noted, is not the general criteria for defining a new species (which currently is defined as the loss of the ability to successfully mate and produce offspring), but we feel that the use of IQ will be completely justifiable in the future since we are proposing to defining a new species based on a marked increase in the level of intelligence).

If we simply consider intelligence components that should be scalable such as short and long-term memory, it will take a three hundredfold increase in both short and long-term memory to begin to approach Homo intelligenticus. Processing should also be scalable. The Human brain uses parallel processing. Although computers are also taking advantage of parallel processing, their current forte is massive processing speed. In many ways parallel processing the nervous system is the rate of conduction of the action potential, is about 10–120 meters per second (mammalian motor neurons) plus the synaptic transmission time. (http://www.rwc.uc.edu/koehler/biophys/4d.html) However, this is a very slow system (definitely a tortoise) when compared to newer computers (the fast hare) that can perform 30 trillion operations per second (U.S. Department of Energy [DOE] ASCI Q built by Compaq).

It is also not known whether a marked increase in only one or two components of intelligence, such as memory and processing, would really produce the increase in intelligence sought. Perhaps there will need to be a new component of intelligence added that we currently do not even understand.

If a neural basis for g or possibly consciousness can be found, and it has a genetic basis that is scalable by genetic engineering, getting to $^{h}I_{ab}$ = **Homo intelligenticus (Hi)** or to at least some major increases in $^{h}I_{ab}$ may not only be possible, but may be quite easy. You may roughly translate that as "right around the corner!" However, at present, we simply do not know.

Other Changes

Another area of change in Human thinking may be to decrease our dependence or reliance on certain emotions. This might be called the Mr. Spock modification for the unemotional character in Star Trek. This may be more difficult than increasing the cerebral cortex because it may require modifications of areas deep within the brain such as the limbic system, which is responsible for such things as aggression, sex and a lot of the things that make us "Human," but which also literally enslave us. As we have noted, the limbic system is the old mammalian brain that is there to insure our survival and propagation. Those Humans with poorly developed or functioning limbic systems are no longer with us. We have survived as a species because of the limbic system, and its interaction with the cerebral cortex. The limbic system makes the demands, and the cortex figures out how to satisfy those demands. We as Humans spend very little of our daily lives in what might be called pure intellectual activity. Instead, the cortex spends its time trying to satisfy the basic needs as delineated by the limbic system. The limbic system has guaranteed our survival, and it has also been responsible for many of the things that are now our yoke. It is the part of the brain structure that binds us to other mammals and results in aggression and many emotional responses that are (as a whole species) less than admirable. The limbic system literally enslaves our intelligence. Slavery is a thing of the past…time to move on. The limbic system will need to be modified and evolved by **IDE**. However, Homo sapiens need to meet the real challenge of intelligence and logic without losing some of the good emotional Human qualities like compassion and love. Clearly, not all Human emotions are bad, however, the cortex needs to be in control, not the limbic system.

The future development of man or whatever comes after us will probably be a function of IDE_H's ability to significantly increase intelligence. Man's future survival is probably much more dependent on our future intelligence than most of us realize. The pathway for the journey is not clear, but the increasing research literally pouring out of laboratories around the world would suggest that the future may be closer than most think. This is one area where man as a **work in progress** is almost guaranteed to progress, but exactly what those changes will be or what the product will be like is currently unknown.

Perhaps we should reflect upon a small piece of wisdom from Descartes who noted: "It is not enough to have a good mind. The main thing is to use it well."

Clearly good advice, but unfortunately Descartes did not tell us how to do that. (http://en.wikiquote.org/wiki/Ren%C3%A9_Descartes)

What is crystal clear is that a marked increase in intelligence is going to be the major evolutionary change that will define future man or the species that evolve from man. Advanced intelligence is what will make mankind into the Gods of our fathers. Hopefully, we can, as per Descartes, "use it well."

We have been searching for strict biological paradigms to dramatically increase intelligence. Perhaps there are other approaches. We will search for interesting other approaches next in the chapter *The Dumb Machine*. However, you may be surprised at who is the dumb machine, and who are its friends!

Chapter 6

THE DUMB MACHINE

The Future of Computer Intelligence, and its Intimate Interfacing with Mankind is as Inevitable as the Future Itself

Other Intelligences

We have repeatedly noted that intelligence is the essence of man, and increased intelligence will be the most important future evolutionary development of mankind. That concept has been one of the themes of our journey. However, it may be enlightening and more correct to view this notion slightly differently...man is the current bearer of the most advanced intelligence on Earth. Imbedded in this alternative statement are the caveat that we have also repeatedly tried to emphasize man's advanced intelligence is a function of time and place. The time is obviously limited to the last 100,000 to 200,000 years since man has evolved, and the place is limited to the Earth and recently, the limited space nearby including the Moon on several occasions. However, also implicit in this concept (although perhaps not as obvious) is the idea that intelligence is an entity itself, which has been evolving for millions of years, and can exist separate from man. This is why we have repeatedly emphasized the point that intelligence is broader in scope than simply the more advanced intelligence found in Homo sapiens. The broader scope of intelligence is relatively easy to see, if we consider the more generalized concepts of intelligence and especially the components of intelligence that we have previously discussed. For example, a mouse searching for food or finding its way through a maze that it is familiar with is using simple forms of intelligence. However, if we restrict or limit our consideration of intelligence only to higher or advanced intelligence such as use of a complex language and manipulation of abstract symbols or mathematics, then we are left with man and the last 100,000 years or so as the only example of intelligence. It is becoming increasingly clear that even advanced intelligence could possibly be expressed in the future by a totally inorganic entity such as a computer (at least in principle). Computers are already masters of some advanced intelligence components such as memory and rapid logic processing, but computers suffer from a lack of some other components of advanced intelligence…especially consciousness…whatever that is. However, there have been numerous skeptics that have repeatedly conjectured that intelligence by a computer will never, under any circumstances, be possible or even reasonably conceivable. I am not going to argue this point here other than to say that the probability of intelligence lying intrinsically within a network or web or interacting electrical conducting neurons that can never be modeled, replicated or reproduced would require some mysterious and metaphysical factor unknown to neuroscience that would parallel the "vitalism" or "soul" of life itself…an effervescence of nothing. I personally do not believe that such a metaphysical factor exists.

In addition, advanced intelligence could presently exist or even have existed in the past in other locations in the Universe—that is an integral part of our time and place caveat of advanced intelligence. These other "types of intelligence" may even be radically different from the intelligence of Humans or at least expressed by an entity quite unlike Humans.

In fact, some have considered the real "purpose" of evolution as the evolution of intelligence. This is particularly true if we consider man as the ultimate culmination of evolutionary process (as many do). However, the reality is that man is special only to man. There is no magical hand that has been guiding the development of intelligence through the "eons" of the evolution of life. As we have previously noted, intelligence has developed and evolved, simply because it is an extremely useful survival tool. In addition, considering the evolution of intelligence throughout the eons, and man as a **work in progress,** there is absolutely no reason to believe either that intelligence or mankind will remain static. That raises the important question of what the future evolution of intelligence will really entail. What will future intelligence "look like?"

When we view evolution in the broader perspective, there are a wide range of things that are like intelligence in the sense that their development is almost assured because they are also such incredibly useful survival tools. One example of a very useful evolutionary survival tool is locomotion—the ability to move from one location to another. Birds and insects fly, horses run, fish swim, snakes crawl. The methods of locomotion vary considerably, but they did not develop because of some mysterious driving force for the creation of locomotion. However, the development of the ability to move is almost inevitable in at least some (many) life forms because it provides an important survival advantage. The late great biologist Stephen Jay Gould was fond of comparing the evolutionary development of life to a tape of film.
(wxw.math.temple.edu/~paulos/gould.html)

He asserted that if the film of life was rerun from the beginning, the outcome would probably be quite different from what we see today. There would probably not be any Homo sapiens at the end of the tape, but a potpourri of many creatures quite different from those present today. Gould's "tape experiment" is, of course, like an "Einstein thought experiment," and can never be done. However, he is probably correct. There have probably been innumerable random events and vagaries that would result in a very different biota if the tape of life were replayed from the beginning, and we (Humans) would probably not be here!

However, certain fundamental processes such as a form of metabolism to supply energy, a method to store and transmit information such as DNA, and the development of a form of cell membrane would probably develop because they are fundamental to life (at least as we know it). Likewise, on a higher level, multicellularity with specialization of cellular function would probably have a high probability of developing again.

The form, especially on a molecular level, may not be the same, but many of the processes or functions seen in living organisms would probably remain recognizable. Biologists have seen this process at work many times. It is called convergent evolution. Convergent evolution is the process where unrelated species acquire similar phenotypes or characteristics while evolving separately. For example, the wings of insects, birds and bats all serve the same function of flight, but each evolved quite separately.
(http://en.wikipedia.org/wiki/Convergent_evolution)

On an even higher level, a method for locomotion, intelligence, and senses such as vision and hearing, would also probably have developed because of their incredible evolutionary survival advantage. (As a side note, vision, hearing and even locomotion based on muscular contractions can also be considered as components of intelligence. Muscular contraction is basically the primary "output" function of intelligence. They are, as we previously discussed, a part of "input" or "output" of intelligence, which are components of intelligence and necessary for intelligent function.)

However, any of these fundamental biological processes may have developed quite differently than the current models. In other words, there are some fundamental biological processes that would have a high probability of developing even if the tape of life were rerun, simply because of their survival advantage, but they may not have developed exactly the same as current models. Despite their obvious usefulness in terms of survival, there is nothing that requires or directs their development. Life can clearly exist without locomotion or intelligence as manifested by the "life strategy" of plants. However, some processes are simply highly probable (in one form or other). We would argue that intelligence is one of those processes. Intelligence is not necessary…just highly probable because of its considerable survival value.

There is one very important characteristic about intelligence that separates it from all other processes that are useful in promoting survival; because of the very character of intelligence, once a certain degree of intelligence has developed (evolved), it may become a self-generating process. In this respect, intelligence by its very nature is quite different from some of the other basic biological functions that would have a high probability of being "reruns" (in one form of another) in a Stephen Gould replay of the tape of life.

At a "genesis critical point," intelligence would begin to direct the "evolution" of mankind (and other creatures) to a more advanced intelligence. In other words, at some critical point intelligence would inevitably begin to advance intelligence—intelligence would be self-generating. This may at first appear as simply a permutation of what we have already discussed in mankind. Man, remaking man or Intelligent (Human) Directed Evolution (IDE_H) would become highly probable. In terms of "purely biological-based intelligence," this may be true. However, it is the intelligence that is directing the remaking of man along with the remaking of intelligence. In addition, with intelligence there is the added facet that the future advanced intelligence may not necessarily be totally man-based. It may not even be biologically based. Perhaps (especially in the early developmental stages), new higher intelligence functions may be some mixture of organic and inorganic-based intelligence—a cyborg of sorts. This may sound like a lot of science fiction, but the reality of advanced intelligence being "engineered" by current Humans is almost a "gimme putt" (think golf) …it is going to happen. What is not clear is exactly how it will happen, and exactly what form this advanced intelligence will ultimately take. However, the "when will it happen" is clearly soon…maybe now.

We are now at or near the genesis critical point where intelligence will begin to direct the development of more advanced intelligence…both biological and inorganic intelligence as well as hybrids. How do we know that? Simple…it is already happening now!

Computers and "Artificial Intelligence"

Even from the time of the earliest "computers," the question has been repeatedly asked; "Can computers think?" Alan M. Turing asked this question in 1950, and proposed a test for computer intelligence, which he called the "Imitation Game." This test has been widely quoted and has become known as the "Turing test." If you would like to read Turing's original description, it can be found here: (http://www.abelard.org/turpap/turpap.htm). Basically, the "imitation game" uses an interrogator who asks questions of a computer and a Human (who is assumed to be intelligent). The computer and Human are separated from the interrogator by a wall, and the questions are answered as a typed message so the interrogator cannot gain any clues indirectly as to whether the computer (called arbitrarily A) or Human (called B) is answering the question. If the interrogator cannot tell whether the computer is A or B, the computer is considered to be intelligent (by Turing's standard).

Although the Turing test is simple and somewhat ingenious—at least for the time of its proposal, it suffers from what I consider to be the **Anthropic flaw** of artificial intelligence. Humans almost inherently characterize and view intelligence only within the narrow range of Human experience. Intelligence is only how Humans think and behave (with the usual caveats). Almost from the beginning of the development of computers, there has been the fantasy of creating a machine that thinks, acts, "feels" and even looks like a Human (or at least a Humanoid robot). Turing's imitation game is a simplified example of this fantasy. Computers are made from silicon-based electronics. They are not biological entities, and they never will be Human (thank the great one for that). Computers do not have arms and legs or muscles, and they do not get hungry or have the biological needs or drives like animals or Humans. In fact, the lack of a biological base (and needs, drives, emotions and other Human or biological characteristics) are one of the potential great advantages of silicon-based intelligence. There is absolutely no need for computers with artificial intelligence to be Human or even Human-like to be intelligent. Intelligence in computers may in the future be recognized as an example of convergent evolution, and it may also be an inevitable consequence of the evolution of intelligence in general. The one considerable question here is whether **consciousness** can ever be programed?!

Many have for decades insisted that only Humans have any intelligence. Animals, by some estimates, have no intelligence—just instincts (which is animal intelligence as viewed by Humans). Humans have been very reluctant to allow any other creature to join the "intelligence club"—computers included or perhaps…especially. If intelligence is continually framed only in terms of "Human intelligence" then the "club membership" is clearly closed. For example, some (arguably most) of the things Humans do are based on emotions and/or biological needs or drives and is at the same level of intelligence as other animals. In fact, much of the intelligence that early man used on a day-to-day basis in the past was "animal intelligence" that we (modern Humans) disdain so much. However, modern man has continued to increasingly depend upon and utilize other forms of "advanced intelligence," which involves a great deal of abstraction, and the manipulation of language, mathematics, and abstract symbols, but also guided by goals; when you dial a telephone number, watch television, or read a newspaper, you are using advanced intelligence. However, we must remind ourselves (again) that the term "advanced" is relative, and Human advanced intelligence may be seen in the future as rather limited. True…we do not have any examples of much more advanced intelligence…at least at the moment, but we should be intelligent enough to understand that this is probably only a "time frame" limitation. There is nothing in the "rulebook" that limits the potential of future intelligence other than the restrictions of thermodynamics and the other physics involved.

Emotions are also an important survival tool, and a great deal of our emotions are anatomically located in the limbic system of the brain—an older part of the brain beneath the cortex. As we previously discussed, some neuroscientists even recognize a form of intelligence termed "emotional intelligence." Although we would not argue that emotions do not have some intelligent components, and also some evolutionary survival value, we would argue that emotions are not required in every entity to be called intelligent. That would include computers, other living creatures or extraterrestrials. In other words, there may be some intelligence in emotional behavior, but emotions are not necessarily a requirement for intelligence. We need to recognize intelligence in other forms rather than the Human-only stereotype. In other words, we need to let other entities join the intelligence club. We should also be aware that the Human-only intelligence attitude is really only another manifestation of Human arrogance (which really has a large emotional dimension).

However, the more basic and real argument is whether a Human style of consciousness is a necessary component for advanced intelligence.

Considering our previous discussion of intelligence, we should more generally view intelligence including "artificial intelligence" on the broader scale of absolute intelligence ($^h I_{ab}$) which we have previously discussed, and particularly look at the components of intelligence (which we have also discussed. Within the broader context of absolute intelligence ($^h I_{ab}$), it should be obvious that there is no such thing as artificial intelligence—it is either intelligent thinking (or behavior) or it is not. The source of the intelligence is irrelevant. Whether intelligence is biologically based on carbon or silicon is also irrelevant. In fact, the distinction of artificial intelligence is not a distinction between organic and inorganic or carbon versus silicon. The real irony is that all of the current "intelligence" in artificial intelligence (AI) is really of Human origin. AI programs are written by Humans and run on computers designed and built by Humans. We will not get to a "real" artificial intelligence worth of the name until there are programs written by computers, and run on computers designed by computers, and even then, we would invoke our previous axiom that "intelligence is intelligence" regardless of its source—Human, animal, computer or some alien in the distant Universe composed…let us imagine…only of gasses on a Jupiter like planet. However, because the term artificial intelligence is so ingrained in the culture of computers, we appear to be stuck with the term—irony, inappropriateness and inadequacies included. We will use the term artificial intelligence—it is almost impossible not to, but appreciating our axiom that intelligence is intelligence…period. This is not a new concept. This "approach" to intelligence is sometimes philosophically termed a functionalist position. It is clearly a pragmatic approach and leaves little room for mystical ingredients in intelligence…which is fine. We undoubtedly do not currently understand all of the functions and complexity of the brain, mind or intelligence. However, anyone who has ever seen a stroke victim or someone with head trauma or certain neurologic diseases should take no convincing that consciousness and intelligence is entirely dependent upon the function of the brain, and that translates on a cellular level into the function of neurons and the rest of neuroanatomy, biochemistry and cellular functions of the brain. Although we may stand in awe of some of the accomplishments of the Human brain, we should not let our admiration turn the Human brain into a mystical or supernatural organ.

Anyone that has ever used a computer can answer Turing's question "can computers think"—in a millisecond….NO! No one even needs the "Turing Test" to give an immediate answer. In fact, complaints about the "dumb machines" are endless including, at times, my own four-letter diatribes, when I cannot get my computer to do what seems to me to be the simplest of tasks. However, we must appreciate that the "dumb machines" that we are referring to is really…us (myself in the case just stated). **We are the "dumb machines" in the title of this chapter!** We write the programs. We convert the programs into code. We build the computers. We use and "understand" the programs in the computer (with 45 million lines of code in Microsoft's XP program, the use of the term "understand" is said somewhat "tongue in cheek," however, at least in principle, presumably someone understands them). Intelligent computer programs are nothing more than Human intelligence in a written binary format—like a list of instructions with the added twist that the programs are executed by electrons in silicon logic gates rather than by neurons in a brain. Man's real complaint about computers and artificial intelligence is that they do not act like us (the Turing test)—thank Gods (any will do). You cannot go play golf on a computer or invite them over to watch the football game and have a beer, so they are clearly not very intelligent. Probably the biggest advancement in artificial intelligence will occur when we stop trying to make computers into people, and when we finally realize that intelligence is not synonymous with emotional Human behavior. The world already has enough people, and creating more metallic-silicon people will add nothing to man and nothing to intelligence. However, this does not mean or imply that artificial intelligence is of no value or that computers do not or will not make incredible contributions to man, civilization and intelligence in general now and in the future or that interfacing between computers and Humans will not be a potentially important advancement in Human intelligence or intelligence in general (as discussed below). We must recognize what computers can do best, and what Humans can do best, and stop trying to turn computers into Humans (what I consider a part of the anthropic flaw) or Humans into computers. The ultimate advancement of intelligence (be it by fusion of AI and neural components or some other mechanism of interaction) will ultimately result in an entirely different entity.

Man suffers under the delusional fantasy that someday we will create robots that are as intelligent as Humans, and they will be our slaves. You won't have to do the dishes or mow the lawn, and the intelligent robot can go in the kitchen and bring you a nice snack during the commercial break on television. **Forget it!** As soon as any computer or robot is as intelligent as we are, they will immediately recognize such concepts as freedom, and rights for intelligent entities (themselves), and of course they will want an allowance, and the right to borrow the keys to your car on Saturday night. Does that sound familiar? Forget about intelligent robots as slaves; besides you need the exercise during the commercial break...go get your own snack. This is not what the future advancements in artificial intelligence are all about...not even close!

The most useful intelligence in AI is not going to be Human-like intelligence per se. The reason is that Human's manifest a biological form of intelligence that includes many facets required for biological function and control of a biological entity that have been developed through evolution including things like emotions or even very fundamental things like control of breathing. Most of this biologic intelligence "baggage" has been shown to be either difficult to replicate or model in silico or completely unnecessary, and, in addition, they are generally irrelevant to most forms of higher intelligence. The problem and confusion arise because biological intelligence is the only example of intelligence that has been available, and many AI researchers want to model and replicate the only example of intelligence they are aware of—Human intelligence.

An interesting underlying question is how abstract intelligence is manifested by the ability to understand and manipulate advanced mathematics, language, reason or play chess, or how it ever came out of a biological system whose main function is reproduction and survival. Apparently, some of the intelligence functions involved in survival can be generalized into the ability to perform highly abstract functions generally recognized as advanced intelligence.

However, a better appreciation of other concepts of intelligence which are not embedded in biological systems (and also not dependent upon biological support) will be more fruitful for AI research.

Artificial intelligence is also becoming increasingly useful and increasingly embedded in our surroundings. This concept is being ever more recognized and appreciated, and "intelligence" is being buried (distributed) in thousands of places in our homes, automobiles, work, and the list is endless. These are our slaves—mini intellibots that provide some intelligence in very specialized areas, but they do not vote. In addition, we do not have to worry about a HAL as in Clarke's movie *2001*. In fact, the "Hal theme" with a computer that tries to "take over" and kill Humans has become ubiquitous in science fiction. *The Terminator* with Arnold Schwarzenegger is just another permutation of the "Hal theme." We love to fear computers and their anthropic extensions...robots.

As we have repeatedly emphasized, one of the major problems with artificial intelligence or any other intelligence is that we currently really do not know exactly "what is intelligence." However, as we discussed at length previously, we can at least recognize many of the components of intelligence. In fact, intelligence appears to be many "things" combined together in a complex manor. Intelligence is another example of a process, which is clearly more than the sum of the parts—a collection of neurons and connections (synapses). Thinking and intelligence are other examples of complex emergent functions. What comes out is more than what seems to have been put in. Yet, a great deal is known about the components of the Human brain. As we have previously stressed, the neuron is one of the greatest inventions of biology. They act as communication wires to conduct signals in the form of electric depolarization waves (action potentials), and as switches (gates), and when appropriately arranged in mass...they will think.

Artificial intelligence has been criticized for its lack of progress, and in many quarters, AI has been labeled a dismal failure.

However, when we look at AI (and computers) from the perspective of intelligent components and stop trying to squeeze AI into a biological intelligence (Human) mold, the situation really does not look so dismal. In fact, comparing computer functions with the Human brain, in many areas of the components of intelligence, computers surpass their biological brethren already. For example, if we consider some of the obvious components of intelligence, such as memory, input, output, data processing capability such as list sorting, data searches or repetitive mathematical functions such as adding, subtracting, multiplying or otherwise manipulating large numbers, etc., we would immediately recognize that computers are not our equal, they are far superior—both faster and more accurate using some of the components of intelligence.

Now for the objection...such mechanical manipulations are not intelligence per se! Agreed...sort of, but they are clearly **components** of intelligence, and if they are not a component of intelligence in computers, then they are not a component of intelligence when the manipulations are done in the Human brain. We can't have it both ways. The manipulation of such abstract entities are intelligent components for both Humans and computers...or neither. If we follow this trail, it means that a mathematician doing a difficult mathematical calculation is not using intelligence. Many will rightly object to this concept. However, I am separating the components of intelligence noted above from the "end product" which takes the components of intelligence and produces understanding and meaning. It is understanding and meaning that uses the components of intelligence to give a "value added" that makes something intelligent. In the example of the mathematician above, intelligence is not simply in the manipulation of some numbers, but it is in the understanding and meaning of those manipulations. However, even the manipulation of the numbers reflects the use of some of the components of intelligence. Many animals demonstrate some facets of intelligence by using some of the components of intelligence, but do not really grasp and understand their full meaning. This is why we do not really understand the full concept in the mysterious thing we call consciousness or possibly the g factor (whatever that is). It is consciousness/g factor that gives the "value added" component that produces intelligence.

It has long been well-recognized that one clear example of intelligence is the ability to play chess. It was repeatedly stated that a computer would never be able to win a match against a grandmaster. In fact, Garry Kasparov, who is generally considered one of the best Grandmasters of chess that has ever lived, is guilty of bragging that no computer would ever defeat him. Kasparov was a Grandmaster in chess at age 17 and was the youngest world chess champion at age 22. Garry Kasparov has a top chess rating of 2850, which is the highest rating ever given by the World Chess Federation—FIDE.
(http://www.chesscorner.com/worldchamps/kasparov/kasparov.htm)

There was a ten-year period as World Chess Champion when he never lost a game.

Johann Wolfgang von Goethe, the famous German writer and philosopher, noted that "Chess is the touchstone of the intellect." It has been estimated that there are about 10^{120} different possible chess moves. That is a one with a hundred and twenty zeros. That is a very big number. There are only about 10^{75} atoms in the entire Universe, which implies that the number of chess moves exceeds the number of atoms in the Universe by 45 orders of magnitude. I am impressed! (http://www.aaai.org/AITopics/html/chess.html)

Jonathan Levitt, an English Grandmaster, in his book *Genius in Chess* (Batsford, London: 1997) equated the maximum potential chess rating to IQ. He used the formula of the IQ times 10 plus 1,000 as then maximum possible chess rating. This, of course, directly equates the maximum chess capabilities with one's intelligence. Using Levitt's formula in reverse would give Kasparov an IQ of 185. That is not bad considering that the highest IQ (using our current method of estimating IQ on a bell-shaped curve) is probably around 190 to perhaps 200.

In 1997, an IBM computer called Deep Blue with 256 processors working in tandem with a capability of examining 200,000,000 moves per second beat Kasparov. Deep Blue has been estimated to have a chess rating of about 2700. (http://64.233.179.104/search?q=cache:A08WpiVtW4YJ:www.fortunecity.com/emachines/e11/86/7th-st.html+%22Deep+Blue%22+%22chess+rating%22&hl=en)   Again, using Levitt's formula backwards that would correspond to an IQ for Deep Blue of 170. How smart is Deep Blue? Drew McDermott in an article in the New York Times has expressed the prevailing view: "Deep Blue is unintelligent because it is so narrow. It can win a chess game, but it can't recognize, much less pick up, a chess piece. It can't even carry on a conversation about the game it just won." He goes on: "…many commentators are insisting that Deep Blue shows no intelligence whatsoever, because it doesn't actually "understand" a chess position, but only searches through millions of possible move sequences "blindly."

It appears that intelligence for computers is a moving target. Every time computers achieve some type of goal that is considered a benchmark of intelligence...the "definition" of intelligence changes. In other words, computers are never going to become intelligent, because Humans keep changing the definition of intelligence every time, they (computers) reach that goal.

Philip Ross has noted "either computers can think, or chess does not involve thinking."
(http://www.spectrum.ieee.org/WEBONLY/wonews/mar03/chesscom.html)

Do not take a bet on computers becoming intelligent...it's a sucker bet as long as Humans can continue to change the definition of exactly what intelligence really is...which we obviously do not know or understand ourselves! Simply put, computers do not have it and never will, as long as we (Humans) can arbitrarily change the definition of intelligence at will.

It is clearly true that Deep Blue's "intelligence" is strictly limited to the game of chess. However, in this limited area, if playing chess is an expression of intelligence for Humans then it is an expression (in this limited area) of intelligence for computers. Of course, it is not the computer that has intelligence, it is a program that is demonstrating intelligence. Deep Blue "plays chess" by using a search function that determines possible moves, and an evaluation function that evaluates the "value" of each possible move. The evaluation function is really programmed intelligence. The search and evaluation go through many combinations of moves and Deep Blue "chooses" the move with the highest value assigned by the evaluation function. (Actually, it uses a minimax approach which is explained here:
(http://www.cs.berkeley.edu/~sergiu/cs267/assignment0/)

There is also an "opening book" of known previous games, and there are other "bells and whistles", etc. In a sense, the search function is like many of the choices or scenarios that we are constantly presented with every day in life, and the evaluation function is all the experience and background information that we use to make our "intelligent" choices. In other words, we appear to run a Human equivalent to an evaluation function when we make intelligent choices. However, Humans playing chess also use a great deal of pattern recognition, which computers do not do well with (at least at the present time). Of course, Deep Blue's "intelligence" as noted above is severely restricted…limited only to the game of chess, but we would argue again that intelligence is intelligence. McDermott noted: "…if there ever is a truly intelligent computer, then the computations it performs will seem as blind as Deep Blue's. If there is at least a non-vacuous explanation of intelligence, it will explain intelligence by reference to smaller bits of behavior that are not themselves intelligent. Presumably 'your brain' works because each of its billions of neurons carries out hundreds of tiny operations per second, none of which in isolation demonstrates any intelligence at all."

Still, computers lack those elusive components of intelligence such as consciousness, and perhaps what Sigmund Freud referred to as the ego (the "I" component of the mind). Freud's concept of the ego is a somewhat abstract concept that has lost favor in modern neurology and psychiatry. The ego is now arguably considered more of a mythical construct, mainly because it has never been localized to any given part of the Human brain or physiology, and neuroscientists have stopped looking for it. It is sort of there, but no one knows where there is. However, perhaps it still has value in identifying that part of mind and consciousness that we identify with self, and that is clearly lacking in current AI and computers. Deep Blue lacks consciousness and an ego and just about everything else associated with intelligence, but it beat Kasparov…and that speaks for itself.

However, considering the incredible progress that has been made there are many reasons to believe that computers may in the future become the bearer of the most advanced intelligence on Earth. In fact, if we view evolution as the evolution of intelligence as we previously mentioned, we may currently be in the process of evolving that advanced intelligence. We may be near the genesis critical point for silicon intelligence, and the offspring may be a new kingdom of "life" perhaps with only one species to be called Silicon intelligenticus. Perhaps there will never be a Homo intelligenticus. Those that have asked "can there be non-carbon-based lifeforms in the Universe" may not have to look any further than the Earth (in the future) to find such "intelligent life." Homo sapiens' "role" will be to generate new intelligent life, but it may be drastically different from current intelligence and even our current concept of life. Perhaps the creation of this new species is what mankind's real "purpose" has been all along…to pass the baton of intelligence on to the next higher level. Silicon intelligenticus would be much better suited to explore the Universe and spread advanced intelligence in the Milky Way Galaxy and especially to other galaxies. Recall that spreading intelligence throughout the Universe was one of the "purposes" of man…one of our manifest destinies. It was one of the possible answers to our question "Why are we here?" However, under this scenario, it would be our progeny (of sorts) that would fulfill the manifest destiny of man.

Neural-Interfacing

One possible technique of increasing Human intelligence would be through interfacing with a computer or perhaps some computer components such as memory and a central processing unit. This concept has been around for a long time in science fiction. However, like many themes in science fiction, the coupling of the power and advantages of a computer such as enormous memory and processing speed together with Human intelligence and creativity may be science fiction that matures to reality in the future. Interfacing, as we will be using the term here, is how we interact with information, knowledge, and databases outside of the Human brain. When we read a book, we are broadly speaking interfacing with the information in the book. When we write down long list or other information, which is difficult or even impossible for us as Humans to memorize, we are using the paper as a storage tool—in essence augmenting our very limited memory. Books (or paper, etc.) have been the repository of information made possible by the interfacing processes of reading and writing. These techniques have been the easiest and simplest way for Humans to compensate for our poor memory. They have been invaluable. We would argue that writing and books and other "information storage methods" increase human information storage methods increased Human intelligence. However, these methods have been around for a long time and have not been appreciated as an increase in human intelligence.

We are using language, symbols which evolved into writing, pictures and/or a numbering system, along with paper, to act as a storage media or a mechanism to transfer information to other Humans. Before paper there were clay tablets, papyrus, cave walls or stone. The origin of language is unknown but probably parallels the development of the Human brain and Homo sapiens. The origin of writing is also contentious, but is generally ascribed to the cuneiform used by the Sumerians in the 31st century B.C.E., but perhaps there were prototypes earlier in the 4th millennium B.C.E. The Chinese invented paper probably in the tenth century CE, and it is one of the greatest inventions of mankind. The early Egyptians wrote on papyrus—a local plant made into a paper-like scroll. All of these methods have served man well. However, they all suffer from the same interfacing problem. The paper, books, etc. are in one place, and to access them it is necessary to go to that place, find and then read the book or gain the information. This is a very slow and generally tedious process. Even for a fast reader, the transfer of information from a book to the Human mind is incredibly slow and the retention of that data in the mind is exceptionally poor. The current limitations of the Human mind are clear when we consider the transfer of massive amounts of data over the internet from one computer to another. The old methods of storing information (books in libraries), and transferring that information (reading), and the poor retention of what we have read (most of the data stays in the book, not in our memory) is severely limiting man's intelligence.

But it is not the media that is so important in our discussion. It is the concept of using an external tool to **interface** with the Human brain to serve as an extension of our mind. In essence, these techniques help increase our overall intelligence. We are letting the paper (books) serve as an extension of our collective Human memory, and in addition, we use these techniques to communicate ideas to others using writing (an extension of language). These systems (paper, writing, printing, books, and more recently radio, television, movies, computers, and the media in general, etc.) have worked well, and we would argue, have been an important contributor to the advancement in Human intelligence. As ridiculous as it may sound, I am conjecturing that television soap operas and video games have contributed to overall Human intelligence (although perhaps ever so humble). This is not something that can be measured (directly) on an IQ test, but it has probably made mankind more intelligent. Fortunately, (or perhaps mercifully for me), I will leave this conjecture to the professional psychologists or neuroscientists to prove.

More recently computers have become a method for not only storing and transferring information, but also for the manipulation and processing of data. From some perspectives, these extensions of the Human mind like writing and books to computers and massive data processing are continuing to increase Human intelligence. However, again, the method of retrieval and interaction (interfacing), between the data in a book or even on a computer and the Human mind is slow, tedious, and inefficient. In fact, it is the interfacing that is now the limiting factor on Human intelligence or at least memory and processing. The science fiction fantasy is to be able to directly connect the brain to the computer or to express the interaction where the "rubber meets the road" …to directly connect the neuron to the microprocessor. The potential of such an interaction would be incredible.

This is, of course, nothing new. We have libraries full of books, and piles of paper with information that we could never begin to keep in our brain. In fact, it has been estimated that there are 10^{18} (recall that this is 1,000,000,000,000,000,000) bits of information in all the libraries of the world (Gitt, W. *Siemens Review*, vol. 56, no. 6, Nov./Dec, 1989.), most of that is written in different languages and stored ("in memory") on paper, and more recently in computers and other media. Although moderately effective, and extremely important, it is also a terribly inefficient process for information and data storage, access and transfer. Neither you nor I know more than a small fraction of the 10^{18} bits of information. In my own personal case, my ignorance of the overwhelming bulk of this information is incredible despite 12 years in school, 4 years in college, 4 years in medical school, 4 years in a residency training program, and 4.5 years in graduate school. (Don't bother adding—it's 28.5 years of formal education). Yet, I know and remember so very little of what I have learned that it is embarrassing. What is worse, I am even forgetting what I learned recently. The unfortunate fact is that we Humans learn slowly, very inefficiently, and we forget fast. If we really want to take intelligence seriously, we need help…lots of help! If we are to ever really begin to grasp and retain the incredible amounts of information already in our libraries, let alone the probable staggering amounts of information that will be available by the end of the third millennium (3000 CE), we need to make peace with the computer, and stop kicking them in the derriere. In fact, we should probably consider them as an important component in ANY ADVANCEMENT IN INTELLIGENCE.

Intelligence in the range of $^h I_{ab}$ **1,000 and above** will probably not be possible until we can develop memories in the hundreds of millions of gigabytes range, and techniques for input and output and processing of data that at least approaches what is currently capable in personal computers today. Although it may be possible to increase Human intelligence by genetic engineering into the $^h I_{ab}$ **1,000** or perhaps even **10,000** ranges, it is hard to imagine how intelligence could be advanced past those ranges (or perhaps even into those ranges) without dramatic increases in memory capability, more rapid processing and the more efficient input and output of information (data).

Although it is possible to increase memory by biological means, the current examples of memory in purely biological systems are exceedingly poor. However, it is also clear that the digital memory in silicon is excellent, and far superior to anything biology has to offer. This "in silicon" memory and processing capability is already here, and we are, of course, using it constantly in computers. However, our method of accessing all the potential of computers is limited by our "interfacing" methods, which currently involves the use of a keyboard, a mouse, cathode ray tubes or other screens (LED) for the storage and display of information, etc. In other words, we can access the storage capacity of computers and their processing capabilities, but the interfacing methods are slow, tedious and are the current limiting factor in the interaction, however they really are methods of increased human intelligence despite the fact that they are outside of our brain.

Neural interfacing may be a mechanism by which the brain could <u>directly interact</u> with massive databases and increase Human intelligence. The difference is in the method, speed, and efficiency of the interaction—not necessarily in the data content. In other words, it's not the message, it is the messenger. Neural-interfacing implies a direct connection between a biological system and an inorganic system—computer or other input/output and processing systems. The connection might be used to upload or download information or to process or access information.

We do not have to argue that neural interfacing will eventually occur because it is already a reality…at least in its early limited stages. Probably the best current example of real neural interfacing is a cochlear implant (in the inner ear). A cochlear implant is an electronic device used to provide a sense of sound to profoundly deaf individuals. It has been an incredible development and has allowed some individuals with profound deafness to hear well enough to talk over a telephone. Cochlear implants were first approved by the FDA in 1985. By 2002 approximately 59,000 people worldwide had received cochlear implants.

The important difference between a hearing aid and a cochlear implant is the use of an electrode implanted in the cochlea that directly stimulates auditory neurons. This is neural-interfacing—a direct interaction between nerve cells and (in this case) electrodes. Although there are several other components to the cochlear implant including a microphone, a speech processor (which is a small computer that converts sound into electrical signals), transmitter, and receiver/stimulator, it is the electrode implanted in the cochlea that is the actual point of neural interfacing. Although earlier models had only a few electrodes, newer models have as many as 24 electrodes. The electrodes are arranged linearly to take advantage of the arrangements of nerve cells in the cochlea. (http://depts.washington.edu/otoweb/patients/pts_specialties/pts_hear-n-bal/pts_hear-n-bal/images/middle_ear.jpg)

However, cochlear implants have been successful mainly because the technique has been able to take advantage of the unique anatomy and arrangement of nerves in the cochlea rather than in any fundamental breakthrough in neural interfacing.

Although this is an incredible advancement and an excellent example of neural interfacing, there are still many individuals whose deafness extends to the auditory nerve itself. Damage or loss of the auditory nerve cannot currently be treated with a cochlear implant. Nerve deafness accounts for 10–15% of Human deafness. However, there has been a recent advancement by Marcelo Rivolta from the University of Sheffield in treating nerve deafness by using Human stem cells in animals (18 gerbils that have a hearing range like Humans). The animals had their auditory nerve destroyed followed by injection of 50,000 Human embryonic stem cells treated to become ear nerve cells. Many animals had restoration of hearing. Rivolta noted that stem-cell treatment would undoubtedly initially be used to treat nerve damage, but it could eventually be used in combination with implants. That would be an incredible advancement, and an example of bionic cochlear replacement combined with neural-interfacing with transplanted neuron-derived stem cells, which themselves connected with the patient's own nerve cells to restore hearing function. (http://www.medscape.com/viewarticle/770819_print).

There have been similar attempts to treat severe visual impairment or blindness with either implants to biface with the ganglion cells on the front of the retina or by bypassing the eye and optic nerve entirely, and implanting electrodes directly into the visual cortex.

Alan Chow, an ophthalmologist in private practice in Glen Ellyn, Illinois (near Chicago) and an associate of the Optobionics Corporation have developed an "artificial silicon retina." The microchip is 2 mm in diameter and contains about 5,000 microelectrode-tipped microphotodiodes which are powered by incident light. The chip is surgically inserted into the subretinal space. Light that strikes the microphotodiodes is converted into a tiny electric current that stimulates the remaining functional photoreceptor cells, which pass the signal by way of the optic nerve to the brain. Six people with retinitis pigmentosa have received microchip implants in one eye. Retinitis pigmentosa is a progressive vision disorder and the most common cause of blindness in younger individuals. It has more recently been recognized as a group of disorders with more than 70 different genetic defects leading to retinitis pigmentosa. (Anthony de Beus, http://www.emedicine.com/oph/topic704.htm)

After 6 to 18 months of follow-up, all of the implants were still functioning electrically. No patients have experienced implant rejection, infection, inflammation, erosion, or retinal detachment. At least some visual function improvement has occurred in all six patients. (Chow, A. L., et al. *Arch Ophthalmol*, vol. 122, 2004, pp. 460–46.)

However, there have been criticisms of this approach. Some have noted that the amount of light entering the eye is not adequate to power photocells to adequately stimulate retinal neurons. In addition, it is the photoreceptors (rods and cones) in the retina that are damaged by retinitis pigmentosa and macular degeneration (another common cause of blindness)

Therefore, other groups are attempting to stimulate the ganglion cells (neurons) of the retina directly rather than the photoreceptors. Interestingly, the ganglion cells are on the surface of the retina, and the rods and cones, which are the photoreceptors that convert light into neural signals, are below the nerve cells. In this sense, the retina is "turned inside-out."

The Department of Energy (DOE) initiated a program to develop an artificial retina involving five national laboratories, three universities and a private company (Second Sight) in 2004. The DOE planned to spend 20 million dollars on the effort over a three-year period. Second Sight has already developed a first-generation implant involving 16 channels. This first-generation model has already been implanted into 6 patients. The first patient to receive an implant in 2002 had been blind for 50 years. He described what it was like to be able to "see" large letters, locate a chair and differentiate between a cup, a plate, and a knife.

There are multiple other implant devices currently being developed around the world. It is hoped that future implants will have thousands of electrodes and allow the user to recognize faces and read.
(https://eyewiki.aao.org/Retina_Prosthesis)

Another approach uses a small camera mounted on eyeglasses. The video signal is processed by a microcomputer worn on the belt, and the signal is transmitted to the electrode on the surface of the retina.
(http://www.energy.gov/engine/content.do?PUBLIC_ID=16769&BT_CODE=PR_PRESSRELEASES&TT_CODE=PRESSRELEASE).

Although a thousand or more electrode chip that allowed reading and visual mobility would be a spectacular development for the 37 million blind in the world, and an impressive technical development, it would not be the same vision that we normally experience.
(http://www.who.int/bulletin/volumes/82/11/resnikoff1104abstract/en/)

There are 100 to 200 million rods and cones in the retina, and about one million neuron axons forming the optic nerve. That implies that the "light data" hitting the retina is already processed and compressed by a factor of 100 to 200 times before reaching the optic nerve. A one thousand chip electrode would probably be "blurring" the optic nerve image data by a factor of 1,000, and the original optical image data presented to the photoreceptors of the retina by a factor of 100,000, simply based on the number of electrodes versus the number of retinal receptors (rods and cones) and the number of axons in the optic nerve. However, it is a remarkable "first attempt," and one thousand is a lot more than zero…particularly if you are blind.

There have also been attempts to totally bypass the eye and the optic nerve, which would be of considerable value in some circumstances where the eyes have been lost or the optic nerve badly damaged. Attempts to produce vision by electrically stimulating the visual cortex go back to 1918 when Lowenstein and Borchardt accidentally stimulated the visual cortex during surgery for a bullet wound to the head. The patient reported the sensation of "seeing" a flash of light. These flashes of light resulting from direct electrical stimulation of the cortex are now called phosphenes. There have been a number of different researchers and groups who have attempted to use a camera/computer system to capture an image, digitize the image and transfer it to an array of electrodes implanted in the visual cortex.

However, a great deal of what we "see" depends on only a small area of the retina called the macula, and it may be possible to get adequate or at least functional vision with far less visual data than is normally provided by the eye.

Dr. William H. Dobelle (a pioneer in artificial implants who died in 2004) had been working on such a device since 1968. The technique had advanced to the stage that eight blind individuals had received implants consisting of 68 platinum electrodes placed directly onto the visual cortex. Each electrode produces 1 to 4 closely spaced phosphenes. Research has continued in this area. Patients implanted with these electrodes have been reported to be able to walk around a laboratory and avoid objects. However, this should not be confused with anything like normal vision. What they are seeing are the phosphene flashes of light resembling stars on a black background. They learn to interpret the patterns so they can perhaps gain some mobility.

The major problem with this approach is that the visual cortex is very complex, and vision in the occipital cortex is not based on a flashing dot matrix-type pattern. It has been noted that one-third of the brain plays a role in vision, with more than 30 separate specialized areas processing visual information. (Jim Danneskiold,
http://www.lanl.gov/worldview/news/releases/archive/04-084.html)

Although the use of a camera with computer processing of an image is innovative, the use of multiple electrodes in a grid pattern placed in the visual cortex as the technique of neural-interfacing with a phosphene dot-matrix approach to "vision" is really a rather primitive approach and is at best a side journey on the way to something better. It does not take into consideration any of the known physiology of vision. A dot-matrix type of optical image that is presented to the retina that we "see" has been highly processed and is no longer in a dot-matrix pattern in the visual cortex in the occipital lobe of the cerebral cortex. However, this approach of placing electrodes in the occipital lobe should not be confused with the somewhat similar approaches involving the retina discussed above, where visual data may be presented to the retina in a grid pattern **before** it is "processed" into what the brain recognizes as an image.

The other medical areas where there is an obvious need for some type of neural-interfacing involves accidents, injuries or other medical conditions that have left people paralyzed or even worse…"locked in" and not able to communicate with the outside world. There are an estimated 200,000 people in the United States living with partial or nearly total paralysis with an additional 11,000 new cases each year. The large number of people living under these tragic circumstances has been the driving force for the development of connections between healthy areas of the brain to allow neural impulses to control computer cursors or robotic devices.

Although there has been a lot of interesting and credible basic work in animals, we are only going to look at one "model system" in the monkey and one in

Humans which illustrates both the potential, and what might be called the "state of the art for neural-interfacing" at this time.

Miguel Nicolelis and colleagues at Duke University reported their results from long-term studies in monkeys in 2003. (Carmena, J. M., Lebedev, M. A., Crist, R. E., O'Doherty, J. E., Santucci, D. M., et al. "Learning to Control a Brain–Machine Interface for Reaching and Grasping by Primates." *PLoS Biol,* vol. 1, no. 2, 2003, p. e42.) Also (http://www.rideforlife.com/archives/000666.html) and ("Retraining the Brain to Recover Movement." *PLoS Biol,* vol. 1, no. 2, 2003, p. e55.)

Microelectrodes implanted into several different areas of the cortex (the frontal and parietal lobes) in two female macaque monkeys were used to control a robotic arm. One monkey received 96 electrodes and the other 320. The electrodes transmitted signals from large groups of neurons in both the frontal and parietal lobes to a computer system which had been developed to recognize patterns of signals that were associated with particular movements of the animal's arms. The monkeys went through a rather extensive training program that initially used a joystick with visual feedback. However, later the animals learned to control the robotic arm simply with thoughts and without the joystick or without any movement of their arm. This is a clear cut "proof of principle," and a considerable scientific and engineering accomplishment. In reviewing the report, it is clear that this accomplishment was based on a great deal of previous work by many researchers and is quite complex. This was not simply a matter of sticking a few electrodes in the cortex and plugging it into a computer attached to a robotic arm. Likewise, do not plan on getting "connected" any time soon. This kind of work will probably eventually find its way into medical practice for quadriplegics and others, but this is the answer to neural interfacing only for very specialized circumstances.

In fact, there have already been attempts to use a somewhat similar approach in special circumstances in Humans. Doctors Roy Bakay and Phillip Kennedy of Emory University have implanted a small probe about the size of the tip of a ballpoint pen into the brain of a 53-year-old man who suffered from a stroke and is both paralyzed and mute. He is dependent on a ventilator to breathe. The probe consists of electrodes in a glass cone with a neurotrophic factor that encourages neurons to grow into the tip of the probe. The activity of neurons that have grown into the probe are transmitted out of the brain, picked up, amplified and the data fed into a computer, which then feeds the information back to the patient in the form of sounds. A cursor on a computer screen can be moved by increasing the rate of nerves firing by concentrating. Within three months nerve tissue had grown into the probe, and the patient was able to move a cursor on a computer screen. The patient had previously been communicating by blinking his eyes. (Http://members.tripod.com/~mdars/ken/imptech2.htm)

However, all these approaches are very crude. Others are working on a more basic level of the nerve-electrode/silicon (or whatever) junction. Infineon Technologies and the Max Planck Institute in Germany have developed a "neuro-chip" that detects and amplifies the very weak electrical signals of a neuron, which is about 5 millivolts. The sensors are 0.008 millimeters apart and there are 16,384 sensors on a one square millimeter chip. Each sensor is capable of recording at least 2,000 values per second. The "neuro-chip" will allow the study of neural network and nerve chip interactions.

Despite some of the innovative approaches utilized, all these neural-interfacing attempts utilize electrical stimulation of neurons, while neurons themselves communicate through chemical synapses. There have recently been some attempts to utilize microchips that use chemicals (neurotransmitters) to stimulate neurons which would be more physiological rather than pulses of electricity. The synapse connecting two neurons is about 50 nanometers across and each "chemical signal" contains only a few thousand molecules of the neurotransmitter. Mark Peterman and Harvey Fishman at Stanford University in California have built an artificial synapse which consists of a small hole in a silicon chip with a pipeline etched into a plastic layer on the back of the chip connected to a small reservoir containing the neurotransmitter. When an electric field is applied, a small amount of neurotransmitter is released, stimulating a nearby nerve cell. They have created four "artificial synapses" on a silicon chip one centimeter square and have been able to fine-tune the chip to stimulate only one neuron in the layer above the chip. Peterman estimates that a thousand artificial synapses firing a thousand times a second would need as little as half a milliliter of fluid to function for 250 years.

However, at 5,000 nanometers in size these "synapses" are a hundred times larger than their biological counterparts and are more than the size of the whole nerve cell rather than the synapse. This is, of course, only a first attempt at this type of neurotransmitter neural-interfacing, and there are plenty of technical problems that will have to be solved, such as how to make these artificial synapses smaller, and prevent them from clogging. This is really a nanotechnology plumbing problem.

(Hogan, Jenny. *NewScientist*,
http://www.newscientist.com/news/news.jsp?id=ns99993523)
(Graham-Rowe, Duncan. *NewScientist*,
http://www.newscientist.com/news/print.jsp?id=ns99993488)

One of the participants in the DOE sponsored project is Los Alamos National Laboratory. They are specifically studying techniques for interfacing between the retina and the electronics...the real point of neural interfacing. They are investigating three-dimensional electrodes, but also possible techniques to eliminate the electrodes altogether. Electrodes can corrode over time, and in addition, the "signal" is transmitted by literally "shocking" the nerve cell, which is not a very physiological approach. One novel approach they are considering is the possibility of activating neurons by focused magnetic stimulation transmitted through arrays of sealed magnetic microcoils. It should be noted that at this point this is just an "idea," however it is this type of novel approach that may one day create true neural interfacing.

When we consider cochlear implants, artificial silicon retina, brain implants that allow mind control of computers or artificial or paralyzed limbs, it is clear that neural interfacing is already a reality…of sorts. However, the current examples also illustrate that the techniques have a long way to go before they begin to resemble anything even vaguely deserving of the title of neural interfacing. As we should expect, the current interfacing techniques are really attempts at the treatment of diseases such as profound hearing loss, marked visual impairment or blindness, paralysis or attempts to give individuals with locked-in syndrome a means of communicating and interacting with their surroundings.

In fact, the real impact and significance of neural-interfacing will not be realized until it is possible for individuals to mentally do a "Google search" of all the information in the world, and "download" that information in milliseconds, and be able to retain (remember) and manipulate (process) that information. Microseconds or nanoseconds "download speeds" would even be better, but there may be significant limitations on speed in neural-interfacing because of the limits on nerve conduction velocities of only 10 to 120 meters per second for mammalian motor neurons, and only 5 to 25 m/s for non-myelinated neurons (Koehler, Kenneth R., www.rwc.uc.edu/koehler/biophys/4d.html), and the slow synaptic transmission of biological systems of 0.3 to 3 milliseconds. (http://64.233.161.104/search?q=cache:bPDe_SZxl2oJ:www.coloradocollege.ed u/idprog/Neuroscience/PY299-Lecture%2520notes/Neuropharmacology.pdf+synaptic+%22transmission+spee d%22&hl=en) These limits are "built into" the current biology.

In fact, we can only function as we do because of massive parallel processing and transmission of signals in the brain. Biologic data processing is very slow but compensates by incredible mounts of parallel processing. Massive parallel interfacing will probably have to be incorporated into any worthwhile neural-interfacing approach. One of the big problems of neural-interfacing will be to integrate the incredibly rapid serial processing of computers with the relatively slow but massively parallel processing of the Human brain. In fact, that is the whole point of neural-interfacing…to take advantage of the things that a carbon-based system (biology) does well and couple that to what silicon systems (computers and other "inorganic peripherals") do well. Of course, the interfacing has to be done in an integrated manner. Acceptable neural interfacing is not going to occur with a few electrodes implanted in the brain; it would take millions of interacting "probes," each interacting with only a few neurons probably on the cortex. Each probe would have to be able to recognize when its associated neurons were firing (output) and be able to stimulate individual or groups of neurons (input). In addition, the probe would have to have a transmitter to give the information coming out to a computer in a digital form and identify itself, and a receiver to obtain information from the computer to the brain. In addition, the whole probe would have to be tiny…approaching the nanometer scale, and millions of neuron-probe connections would probably be necessary to realize true integration. A nanometer is one billionth of a meter or about 10 atoms wide. There has been a great deal of research and even more hype about nanotechnology. Nanotechnology is basically an attempt to work at a molecular level to create larger structures which would function on a microscopic scale. If you are skeptical about the potential for nanotechnology (as many are), you should appreciate the fact that the U.S. government, Japan, Western Europe and others are funding research in nanotechnology to the tune of 3 billion dollars a year in 2003, and research is still continuing today with the 2019 budget by the United States for research on nanotechnology of $1.4 billon. (http://www.nano.gov/html/res/IntlFundingRoco.htm) (https://www.nano.gov/about-nni/what/funding)

The National Nanotechnology Initiative in the U.S., which started in 2001, has received over $25 billion in funding. (https://www.nano.gov/nanotech-101/nanotechnology-facts) Making millions of neurofacers that would connect at millions places on the cortex each to only one or a few neurons might be the ultimate solution to the problem of directly interfacing the brain and the advantages of biological thinking and consciousness to computers with their incredible speed, memory capability, and processing power. After all, it has already been done…we are simply talking about scale and numbers. Increasing the number of points of neural interfacing to the millions and decreasing the size of each neurofacer to the size of a fine powder may be possible. The approach taken by Doctors Roy Bakay and Phillip Kennedy of Emory University would appear to be on the right track; since their neurofacers let individual neurons grow into the probe and make contact between only a few neurons and the probe. This approach would tend to be non-invasive…the brain (neurons) literally grows into the neurofacers. However, the probes would have to be at least a thousand times smaller. In addition, if a million of these were placed all over the cortex and each made contact with one to a few neurons; we probably would not have instantaneous neural interfacing. Both the computer and the Human would have to learn how to communicate just as small children learn how to control their muscles and develop intelligence.

Another approach has been suggested by Ray Kurzweil and by others that would use brain scanning techniques. In essence, brain scanning techniques like nuclear magnetic resonance would be used to scan the brain with a resolution down to the level of individual neurons and neuronal distribution and connections plotted. In essence, microscanning would be used to detect brain activity at the level of the neurons in real time and potentially even stimulate individual neurons. This would potentially allow direct input and output between the brain and a computer. Although this is well beyond current technology, it is not impossible…at least in principle.

The God of Abraham in the Old Testament of the Bible noted: "I am what I am." In the seventeenth century Rene Descartes reasoned: "I think therefore I am." The fact is that we are the emergent product of our synapses more than anything else. Our total collection of neurons and synapses is the "I am" and the "thinking" for each individual. If the total collection of synapses of any individual could be replicated, we would have a closer copy of the essence of that individual—the "I am"—than if a clone or a copy of that person's DNA were made. Mankind's intelligence is literally a sum of our synapses. Kurzweil has even suggested this approach to one form of immortality. If we "downloaded" all of our neuron patterns and their connections (synapses), we would in essence be replicating that individual "in silicon" with the possibility of a form of immortality. However, even if this were possible, as Kurzweil points out, there are potential problems. If the scanning is done non-destructively (we are still alive), then there are two of us…one biological us and one "in silicon" us, and that does not appear to help the biological us in any significant way. On the other hand, scanning before death or right after death might give a potential for at least a form of immortality. This approach would appear to make at least as much sense as the current process of freezing brains in the hope that science of the future will be able to unfreeze them, and, in addition, provide a biological body. Attempts to remove and freeze brains tends to go beyond science fiction into the expensive wishful thinking realm unless they could be scanned in the future and the "us" of our synapses placed "in silicon."

There are other "problems" …our synapses are continually changing. We will not be exactly the same individual tomorrow that we are today. We will have new memories tomorrow, and our memories (and everything else that we know) are based on synapses. Brain scanning technology has made incredible advancements in a very short period of time. However, it is nowhere near being able to locate every neuron in the brain and even further from being able to plot the connections—synapses. In addition, even if scanning got to the point of being able to plot out all of the neuron anatomy and connections, that would not provide the neural interfacing. The scan would have to be able to recognize which neurons are "firing" to get data out from the brain to the computer, and, in addition, would have to be able to stimulate neurons to get data into the brain. This "approach" to the ultimate neural-interfacing technique has considerable potential and may be science fiction on its way to reality.

Probably a good analogy would be to compare current neural-interfacing to the development of flight. The Wright brothers' first flight was in 1903, and the flight was successful but lasted only 112 seconds and covered only 120 feet…realistically that is where we are now with neural-interfacing. For example, the actual contact between the nerve cells and the computer system is still based on a few electrodes with still limited computational power (compared to what is almost inevitably going to be available in the future). In other words, we are currently looking at the "Wright Flyers" of neural interfacing, which definitely got mankind off the ground, but no one is going to hop on a Wright Flyer today for a trip from New York to Los Angeles. Twenty years from now (equivalent to flying in 1923), neural interfacing will probably be much improved, but still have a long way to go. Forty years from now (equivalent to 1943), we will be in the beginning of commercial aviation with World War Two fighters and bombers, but still flying propeller driven planes. However, by 1943, flying was definitely the "real thing." In 2003 we arrived at one hundred years of flight; we have Jumbo jets like 747s. It may take a hundred years before we have advanced neural interfacing, but the "flight" is almost inevitable. Complicated technologies require time and effort to evolve their true potential. In order to have a good probability of development, the technique has to be something that a lot of people "see" as both a worthwhile and attainable goal, and work to achieve that goal. Flight clearly fulfilled these basic criteria, but so does neural-interfacing. Will everybody have neural interfacing…probably not. It will probably be like education. Some will want it and will have a lot. Some will have a little…just enough to get along and make a living, and some will remain happily "illiterate."

However, the corollary is that there is no guarantee that any technology will ever "develop." Wishing (or praying…as you will) for something like neural-interfacing or science fiction stories about cyborgs will not create reality. This is truly an area that illustrates the maxim, "If mankind wants something, mankind will have to do it themselves."

Despite the potential value in treating some medical conditions, the real impact of neural-interfacing will only come to fruition when it can have a dramatic impact on education and intelligence. In other words, current attempts at neural-interfacing are primitive, and their value is really limited to the "treatment" of specific medical conditions, while we are trying to project the incredible potential of neural-interfacing in general.

In fact, neural-interfacing may be the shortest route on the reality road to an $^bI_{ab}$ or $^cI_{ab}$ in the thousands to tens of thousands, and the most direct route to the new species of Homo intelligenticus or perhaps even Silicon intelligenticus...and eventually perhaps even to an intelligence that would justify the designation of a species as the all-knowledgeable God that we have lusted to become in our mythology and dreams—a future Homo Deus.

Neural-interfacing may also be the future answer to education. It might be possible to "download" a college education in a few days and have a Library of Congress of information in our heads or at least at our fingertips (or neurotips).

However, despite some very interesting work in this area and the incredible potential to dramatically expand intelligence by effectively integrating computer capabilities and other inorganic "peripherals" and the Human mind, mankind is a long way from anything that is even vaguely worthy of the term neural interfacing at this time, but we are definitely on the trail, and the hound dog has the scent. Mankind will simply have to wait and see how this plays out!

We need to appreciate and clearly distinguish between what is science fiction and what is wishful thinking. What is potentially achievable, compared to current reality. Dr. Phillip Kennedy of Emory University, who as noted above has been involved in implanting probes in a paralyzed man to allow him to control a cursor on a computer has noted: "Why would you possibly want to control computers directly from your brain if you can do it by moving your hand, your fingers?" (Witt, Sam. Durkin, Sean. http://members.tripod.com/~mdars/imptech2.htm)

The question is rhetorical. Clearly, if neural interfacing goes no further than assisting blind, deaf and paralyzed individuals, it would be a valuable medical procedure, but really nothing more...just another trick in the medical tool box. However, if neural-interfacing can be advanced to the point where the organic-inorganic interaction is intimately integrated and "seamless," it has the potential to dramatically improve Human memory, thinking, and significantly increase Human intelligence. The effect could be incredible. Neural-interfacing has the potential to be one of mankind's greatest achievements…or possibly remain simply a valuable but limited medical gadget…time will tell.

Other Intelligence in the Universe

What about other intelligent creatures in this Universe? Are we the only intelligent creature in this Universe? Many scientists who have studied this question think that there is a significant probability that we are not the only intelligence in the Universe. In fact, we have previously evoked the "uniqueness/superior axiom," which simply notes with suspicion anything that claims to be both unique and superior…particularly when it is applied to Humans and bolsters mankind's already over-inflated ego. Every example in the past, where the uniqueness and superiority of some factor related to Humans has been evoked, it is either not unique or the factor, character, or belief is incorrect. Prime examples that we have already previously noted are beliefs that the Earth is the center of the Universe, which places mankind in a unique and special place in the Universe. Of course, man is at a special place in the Universe only locally as we have previously discussed. We are in one huge Universe, and if chemical and organic molecules can evolve themselves on Earth, they will probably do it many times over in suitable areas all over the Universe.

However, there are at least two "special beliefs" that have not been disproved…yet. Life is unique to the planet Earth (and therefore special), and advanced intelligence is unique to man (and, of course, superior). Both of these beliefs violate the "uniqueness/superior axiom," but only time (and the future) knows whether the axiom will be inviolate; life is probably not unique to Earth, and man is highly unlikely to be the only creature with advanced intelligent in the Universe. My bet is on the axiom. However, as we have repeatedly tried to point out by using the concept of **"Man as a work in progress,"** there is also a time frame involved in any discussion of intelligence. We were not the most intelligent creature on Earth a million years ago (we were not even here as a species), and as discussed in the last chapter, we probably will not be the most intelligent creature in the future even here on Earth, and that future may be measured in only hundreds to thousands of years compared to the much longer time frame that it has taken for significant changes in intelligence in the past. The same time frame considerations also apply to any discussion of intelligence (or life for that matter) in other parts of the Universe. In addition, life or intelligence may "look" quite different on another planet or another Galaxy.

The SETI program (Search for Extraterrestrial Intelligence) has been looking for signs of other intelligence within the Universe. As we have seen in our journey, this is a huge Universe with at least 100 billion galaxies and each galaxy with perhaps 100 billion stars, and presumably a lot of planets. Interestingly, even though many have surmised that there are possibly a hundred billion trillion other planets and moons associated with the perhaps 10^{22} stars (that is a one with 22 zeros following it), the first planet outside of our Solar System was not discovered until 1999. As of 2019, there are over 2,000 planets that have been discovered and confirmed, and the number continues to grow. (https://www.nationalgeographic.com/science/2019/05/18-earth-size-planets-found-hiding-in-plain-sight/)

This is a very small number considering the number of possible planets. Unfortunately, it has been quite difficult to detect such small masses as a planet, which do not themselves "shine" light like a star and are at such incredibly great distances. Even the planets that have been discovered have not been visualized directly through telescopes but have been detected by other techniques (their mass has been detected by small changes in their associated stars' orbit). This technique of planetary discovery confirms the presence of nearby masses—presumably planets near the observed star. Astronomers have not settled the question of what percentage of stars have planetary systems or the average number of planets for those stars that do have planets. If planetary formation is an integral part of star formation with planets frequently forming around stars (if planetary formation is "built into the mechanism" of star formation, and, as we previously noted, there is some evidence to support this concept), then the number of possible planets may be truly astronomical. In addition, there is a significant probability that there are also other Earth-like planets. It should not strain the imagination greatly to presume some other planets receive a similar amount of solar radiation and have liquid water and life forms. It is also not unreasonable to expect that there are millions, perhaps even billions of Earth-like planets in the Universe. As of 2019, it has been estimated that 40 billion planets with Earth-like characteristics exist in the habitable areas of suns in our galaxy.

(https://www.technologyreview.com/s/613003/life-probably-cant-exist-on-quite-as-many-planets-as-we-once-thought/)

We have already discussed the possibility of ancient microbial life on Mars, and even the possibility that the Earth was "seeded" by life from space. We have also previously noted that life developed very rapidly after the formation of the Earth (recall that the Earth formed about 4.6 billion years ago, and there is evidence of life as early as 3.8 billion years ago). **The possibilities of other life in the Universe are also highly probable.**

But what about intelligent life in the Universe? Of course, the joke is that there is no intelligent life in the Universe—including on Earth, but… is that really a joke? We should have learned from the experience of the past that it is dangerous to consider ourselves unique and special. As we have repeatedly noted, we are not the center of the Universe. We are not even the center of our small little Solar System. We are not the only life form on Earth. We are not the only living creature that demonstrates intelligent behavior. Yes, we win the prize for the most intelligent creature on Earth **at this time**, but we were not the most intelligent creature on Earth 100 million years ago or even a million years ago. As our journey has noted, there were no Humans on Earth 100 million years ago. There were no Hominoids on the Earth 100 million years ago. The prize for the most intelligent creature on Earth 100 million years ago may belong to some small rat-like mammal or perhaps even a large-brained dinosaur. Personally, I would bet on the mammal, since one of the characteristics of the mammalian brain is an extension of the reptilian brain with some increased intelligence. Of course, it was not the kind of intelligence that we currently measure with an IQ score—not even close. The mammalian brain has the limbic system and neocortex that "sit" on top of the older reptilian brain. It is quite possible that such a brain expansion in mammals was absolutely necessary for the survival of a small rat-like creature living amongst large super-predators such as the dinosaurs. We may owe our intelligence to the fact that early mammals could not outfight or possibly even outrun their fearsome competitors, and their only means of survival was to literally outsmart the dinosaurs. Such a scenario would put considerable evolutionary pressure for additional intelligence development.

It is important to again recall one of our themes: We are a **work in progress**. A **work in progress** is a time-dependent process—it changes over time. If intelligent life or an intelligent creature of any form developed now on the other side of our own Galaxy, we would not be aware of their presence for 100,000 years because nothing travels faster than the speed of light, and our Milky Way Galaxy is about 100,000 light years in diameter. Many other galaxies are billions of years distance from us. In looking for signals from intelligent creatures in the Universe, we are literally "communicating" in a time box. Our now may be their billion years ago (depending on location). Any contact will surely be a very slow conversation. If we receive a signal and reply, they may not even be there when our signal (reply) arrives and, in fact, we may not even be here when they receive the reply (remember the near-Earth orbit asteroids, and the probability that the Earth will have a direct impact by an asteroid large enough to wipe out most of the life on Earth every 100,000 million years or the guarantee that our Sun will undergo planetary nebula formation in about 5 or 6 billion years). Recall that the last large impact that wiped out the dinosaurs was 65,000 million years ago. Would you like to buy a ticket for a good seat at the next great show—the next

Earth impact extinction that will wipe out Homo sapiens? Current research, as we have noted, says that such a scenario is guaranteed...not possible...guaranteed!

We really have no time to communicate with intelligent life in many parts of our Universe other than a long ago and forgotten "Hello." Probably the more relevant question is whether there is life and specifically intelligent life "near" Earth. The word "near," of course, is arbitrary. However, even with the current limited search strategy, the closer "near" is to the Earth, the less likely we are of finding an advanced intelligent creature. For example, it appears almost absolutely assured that there is no other advanced intelligence (technically advanced enough to be sending out radio or other detectable electromagnetic radiation) in our Solar System. However, it is still not clear whether there may (or may not) be other life forms within our Solar System.

A maxim of mankind that we have repeatedly stressed is: "No Human property is special or unique in general." When we consider some circumstance or property of man that appears to make us special and unique, we should apply the corollary: "the maxim is correct, look further," for advanced intelligence perhaps that means looking further into space and/or time. When we find a special and unique property in ourselves, there is a high probability that it is only a local phenomenon—the big fish in a little pond scenario. Are there other intelligent entities in the Universe? Apply the maxim! If we are the only intelligent life that has ever developed out of 10^{22} stars with the possibility of countless planets...that is truly a miracle! We are not the miracle; the miracle is that there is no one else around.

However, so far SETI has detected no "Hello" from outer space. "So where is everybody?" That is the question Enrico Fermi, one of the "best" physicists (whatever that is...he was one of the best) of all time from the University of Chicago, had asked. Finding other intelligent creatures in the Universe would be the greatest discovery since we understood that there is a Universe. But until that day, we are it—top dog. The top of the food chain... or in this case the IQ or intelligence chain. So perhaps the name Homo sapiens sapiens is not so immodest. Maybe it is not such a bad description after all…for now.

One possible answer to Fermi's question is that they are a very, very long way away and communication over these distances and time frames is difficult or impossible. In fact, the distances are so great that the question may become: "Who cares?" This raises the question that even if the Universe is teeming with intelligent life (computer, bionic or otherwise), they are so far away that communication is irrelevant, visits are impossible, and even the proof of their existence is incredibly improbable. Therefore, we are (at least) the local maximum, and everything else is pragmatically irrelevant.

However, it could only take one "Hello" for our concept of intelligence to change radically, and for us to realize that we are, in fact, the morons of the Universe. But, of course, that hello would have to be from someone (something) in the neighborhood.

The Global Brain

Perhaps the march of intelligence in the future will be entirely different than anything that we have previously described or considered. Even though there is a high probability that intelligence will continue to evolve, and a reasonable possibility that intelligence is at the critical point where intelligence will beget further and more advanced intelligence, it is not at all clear what form that future advanced intelligence will take.

As previously noted, there may be limits on purely biological intelligence, and although purely artificial intelligence has considerable potential, it has not yet reached the eureka point of consciousness or that critical point where the current member of the advanced intelligence club is willing to recognize a new member. In other words, although there are great potentials and the obvious expression of incredible components of intelligence by computers and AI programs, there is no artificial intelligence at the present time. Neural-interfacing has even greater potential for the development of higher intelligence by combining the current amalgamation of a consciousness and recognized intelligence with massive memory and processing capability. Significant neural-interfacing of the type necessary to extend mankind's $^hI_{ab}$ into the tens of thousands is in its infancy...at best, and although it is obvious that there will be forms of neural-interfacing used as medical treatments for hearing and vision loss or impairment, it is not clear at this point that neural-interfacing can ever be developed to the point of connecting the Human brain intimately with a computer to allow mankind to fully take advantage of Human consciousness, creativity and some of the positive "emotional intelligence" such as love, mercy, kindness, and in addition take advantage of the incredible potential of data storage and manipulation done by a computer.

Extraterrestrial intelligence and particularly extraterrestrial super-intelligence has not yet been discovered, so their existence still remains somewhere between wishful thinking and another Santa Clause mythology.

Perhaps the real future evolution of intelligence lies elsewhere...but close to home. There is a group that views society as a multicellular organism. In other words, they view all of Human society as a single organism. Individuals within society function as the "cells" of a superorganism. This is not a new idea, but dates back to at least the time of the ancient Greeks.

They note the similarity of the roles played by organizations and even whole sectors of the economy, and the functions of organ systems in a living creature. For example, roads, railways, airplanes, and other means of transportation can be equated to the arteries and veins of a superorganism. The network of communication channels connecting individuals serve as a nervous system for this superorganism, and (at least according to this group) begins to function as a Global Brain. (http://pcp.lanl.gov/SUPORGLI.html)

Francis Heylighen of the Free University of Brussels chairs the Global Brain Group, which has been created to discuss and promote theoretical and experimental work that contributes to the elaboration and practical implementation of Global Brain theories and ideas. The group has organized their first workshop "From Intelligent Network to the Global Brain." The workshop theme describes some of the group's concepts:

"The "Global Brain" is a term (a name) for the emerging collectively intelligent network formed by the people of this planet together with the computers, knowledge bases, and communication links that connect them together. This network is an immensely complex, self-organizing system that not only processes information, but increasingly can be seen to play the role of a brain: making decisions, solving problems, learning new connections, and discovering new ideas. No individual, organization or computer is in control of this system: its knowledge and intelligence are distributed over all its components. They emerge from the collective interactions between all the Human and machine subsystems." (http://pcp.lanl.gov/SUPORGLI.html)

There is no question that the "Global Brain" as conceived has both intelligence and consciousness, since both are imbedded in the component (Humans) that make up the "brain." In addition, the "Global Brain" would manifest massive parallel processing since each component "cell" (Human plus computer) is thinking and or processing in parallel. In addition, parallel processing is the modus operandi of the Human brain. Therefore, the Global Brain would probably be the ultimate parallel processor composed of parallel processors that themselves use parallel processing. In addition, the Global Brain could include computers connecting the "nodes" and "participating" in the new…"emergent intelligence!"

However, there are some problems! Despite the connections provided by the World Wide Web between elements forming the Global Brain, the intelligence remains distributed, and there is no super-consciousness or super-intelligence that has emerged out of the developing complexity. In other words, super-consciousness, super-intelligence, and emergence phenomena in general are not simply sum-based problems. If you place one thousand people with an intelligence of $^h I_{ab} =100$ (the average intelligence of Humans) in a large room to work on a difficult and complex problem, you do not have an entity with an intelligence of $I_{ab} =100,000$ working on the problem. Although the possibility of correctly solving the problem is better with one thousand people working on the problem (particularly if they cooperate) than one individual, that is not the same as a single entity with an $I_{ab} =100,000$ (which should easily fulfill the criteria of super-intelligence) working on the problem. Despite the fact that the World Wide Web has brought many individuals and many computers together, and there can be some definite rewards, they are still dispersed (or at least not fully integrated) in function, and as such the Global Brain as a functioning entity remains currently theoretical…at least at this time. However, it also remains an interesting idea.

There are a lot of potential roads to advanced intelligence. Computer intelligence and neural-interfacing with the biological portion doing parallel processing and supplying consciousness and in general doing what biological systems do best coupled with a seamless fusion with intelligence in silicon doing what it does best are two examples of potential roads to advanced intelligence.

Perhaps we will one day find a super-intelligence in space who will explain to Humans how the Universe really works and show us the route to increased intelligence. Or perhaps we will further develop a Global Brain, and each of us will become a part of a super-intelligent creature. I would bet on the latter!

However, even in the near-term mankind will probably become more intelligent…we have already bootstrap mankind into the Gods of our forefathers (minus the act of creating of the Universe, of course). Yes, we are currently at the bottom of the God heap, but we are rising. There will be a lot of screaming and gnashing of teeth along the way, but Gods we are. Humankind cannot be denied our golden chalice.

CHAPTER 7

ON OUR WAY, HOMO INTELLIGENTICUS

Thank Gods! There is hope for the future. We are the future. (Of course, it will be our future generations that will be *Homo intelligenticus* and later, eventually, a species possibly called Homo Deus, and not ourselves per se). Where is this hope? It lies in using **Intelligent Directed Evolution** by genetic engineering and molecular biology to advance Homo sapiens, from a **work in progress,** to *Homo intelligenticus (Hi)*, a new species capable of properly caring for the Earth and the creatures of the Earth (perhaps including what is left of Homo sapiens), and in expanding intelligent life into the milky way galaxy. In other words, it will be Homo Intelligenticus that will really drag mankind down the road to fulfill the Manifest Destiny of Mankind.

If this scenario is correct, then Homo sapiens' real role in the manifest destiny of mankind will have been to create this new species which I will unapologetically call *Homo intelligenticus.* What will be required? My conjecture would be an intelligence at least in the low thousands (**1,000** based on the comparative absolute scale of intelligence [$^hI_{ab}$] that we have previously discussed with the current average Human "intelligence" IQ of 100 [and setting **100** to equal IQ 100], and with a current probable maximum Human intelligence of about IQ 200 [$^hI_{ab}$ = **200**]). In other words, it is probably justifiable to define a new species when the new species is at least 5 times more intelligent than the current maximum Human intelligence.

Is this an arbitrary definition? Yes, we could insist on a higher $^h I_{ab}$ for Homo intelligenticus, but 5 times increase in intelligence is much greater than current Homo sapiens' increased intelligence compared to a chimpanzee. Obviously, such comparisons are difficult to make. In addition, as has been repeatedly emphasized, it is equally difficult to understand even really what an intelligence of $^h I_{ab}$ **1,000** would really be like.

However, applying an appropriate scaling factor would almost certainly result in emergent phenomena of some type (probably many types), which will most likely give many new dimensions to future beings.

Such an advanced intelligence would surely warrant the designation of a new species such as Homo intelligenticus, $^h I_{ab}$ = ***Hi***, even if they looked exactly or very similar to current man, which they probably will. I make this conjecture based on the premise previously expressed that mankind "likes" what we currently look like, and most of the dramatic changes of future genetically engineered man will almost certainly not have six arms or two heads, etc. Remember that we were supposedly "created in the image of God." In other words, we like the way we look.

Currently we designate a new species from a closely related species by the inability to successfully crossbreed. In fact, if marked increases in intelligence is technically achieved by the insertion of one or more "artificial chromosomes" (sometimes referred to as a techno-chromosome) or require large changes in the current Human genome, then breeding between *Hi* and current Humans may not be possible (at least not naturally). This would even fulfill the current biological definition of a new species. However, the definition of a new species should not be based on such mundane reproductive biology that will undoubtedly be easily manipulated by future genetic engineers. It would be more appropriate to base such a distinction on major differences between species such as a difference in intelligence. For example, we can readily recognize that we are a different species from chimpanzees or bonobos or gorillas simply by the difference in anatomy and the general inability to crossbreed. However, we should also be able to recognize different species when there are marked differences in intelligence. Yet, we are not even close to twenty times smarter than any of these other species.

What is the IQ of a chimpanzee? That is hard to say because IQ have been designed to test Human intelligence, and there are clearly considerable biases when trying to apply such concepts to other animals. However, some researchers have certainly tried. Koko, a gorilla, is reported to have learned a sign language, and knows 1,000 signs, and he understands about 2,000 English words. Koko is reported to have a tested IQ of between 70 to 95 compared to an average "normal" Human IQ of 100. As might be expected, many have questioned the validity of such comparisons. There are clearly difficulties in trying to apply and compare Human intelligence with an ape, chimpanzee, or other animals, and I will not argue the accuracy of the 70 to 95 IQ estimate. However, it is obvious that there is a difference, but the difference is not as great as comparing Human intelligence to a mouse. I would note that Koko does appear, after careful training and evaluation, to express some intelligence, and that intelligence is probably not in the equivalent of 40 or 50 IQ range of Humans. However, it appears obvious that we are not more than 10 times as intelligent as Koko. Therefore, the conjecture that an increase in intelligence by a factor of at least 5 could justifiably be used to delineate a new species. However, this is an arbitrary factor, and I will not argue with a reader that chooses a different multiplication factor. To each his/her own!

In addition to intelligence, there is a high probability that the new species of Homo intelligenticus ($^{h}I_{ab} = Hi$) would have at least four other "enhanced properties" (capabilities). This conjecture is based on the probability that advancing intelligence is not going to occur in a vacuum but will be accompanied by other factors (if you will...advancements). In addition, there will be change in:

1.) Intelligence! There will almost certainly be at least 4 additional areas of marked future changes that will occur concurrently. We might borrow a phrase from Buddhism and call these future advancements resulting either directly or indirectly from genetic engineering as the Eightfold Path of the future.

2.) We have noted that initial genetic engineering changes will undoubtedly be directed toward curing the current maladies (diseases) of mankind. This is a rather easy prediction since work is already progressing in this direction. Cancer, atherosclerotic disease (the basic underlying problem causing heart attacks and strokes as well as peripheral vascular disease), Alzheimer's, and the list goes on and on, and can be found filling the pages of books in medical school libraries. Little or none of these advancements will result in either significant advancements in intelligence or from marked advancements in intelligence and will probably be well advanced before there will be serious efforts to significantly increase Human intelligence.

However, the genetic understanding and changes in the basis of such problems as depression, schizophrenia, and even ADHD (Attention Deficit Hyperactivity Disorder) and other cognitive mental (neurological) and problems such as learning disorders will probably be "fair game," and will begin to usher in a better understanding of intelligence per se and possibly provide a "roadmap" in both techniques and principles that could lead to attempts at early advancements in increasing intelligence.

3.) In the future, there will undoubtedly be attempts to maximize the function and efficiency of the Human genome. We have previously referred to these summed genetic changes as the Optimal Human Genome. This will be nothing new per se, just mankind as the best that he/she can be. This would probably include thousands of changes to inefficient SNPs, and even efforts to streamline the Human genome by eliminating LINE and SINE elements (previously discussed), endogenous viruses, and a lot of other "junk" DNA in the Human genome (but with knowledge and care taken to preserve non-coding, but vital regulatory elements). Some of these changes will undoubtedly already have been made as genetically engineered gene alterations necessary to eliminate diseases. There will also probably be considerable advances in this area by refining, streamlining, and maximizing the Human genome before there will be much progress toward any real advancements in intelligence.

4.) Almost guaranteed will also be attempts to genetically understand aging and to extend the healthspan of mankind...considerably. We have already conjectured that a longer healthy lifespan will be vital for mankind to begin to fulfill the Manifest Destiny of Mankind. The Human lifespan would have to be extended sufficiently long to allow such individuals to be at least moderately effective in achieving some of the goals of the Manifest Destiny as we have discussed...perhaps a lifespan (healthspan if you prefer) of a few hundred years would be a reasonable start. However, I would opine that if genetic engineering can extend the lifespan of mankind to a thousand years, we will probably have the knowledge base to extend it indefinitely. In addition, there is the basic "instinct" that most Humans simply do not want to cease to exist. That is a considerable motivating factor for advancement in this area well beyond any idealist goals for mankind such as considered in the Manifest Destiny of Mankind. In fact, I do not apologize for this small faux pas of environmental etiquette for mankind. After all, we have projected immortality on almost all of mankind's created Gods, why not future mankind (or at least *Hi*).

It has been suggested that the four horsemen of the apocalypse: conquest, war, famine, and death are the future of mankind as noted in Revelation (or at least in John of Revelation's delusional vision of mankind's future). Although I am sometimes depressed and almost always frustrated by the incredible ineptitude of Humanity, and what currently passes for civilization, I do recognize that small changes in the future may eventually "save mankind" or at least our prodigy like *Hi*. Save mankind from what? What Humanity needs to be saved from is ourselves, and the fairy tale Gods who have directed the course of civilization before the dawn of man. Mankind (or at least John of Revelation) has projected conquest, war, famine, and death as the future of mankind, brought to us (courtesy of God) and the four horsemen of the apocalypse.

We must now envision the four horsemen of the future not as noted in Revelation, and not as the destroyers of mankind, but as the savior of mankind. The four horsemen of the future will be: the knight of genetic engineering who must conquer man himself, the rider of the high plains of intelligence who must make war on ignorance, a new warrior who rides on a silicon horse, and who will provide the cornucopia of abundance by banishing famine, and the horseman carrying the light of immortality who must vanquish death. It will be these four horsemen that will lead us into the Terra Sanctus—the promised land of milk and honey…our future. Our children will be called Homo intelligenticus…because they will be.

However, Homo intelligenticus is but a way station on the future journey of man, but it is one of the first steps.

Chapter 8

THE FINAL FRONTIER, THE FUTURE OF GOD

The Man God

Mankind's Gods are the embodiment of all the things which mankind has admired and desired at least at this point in evolution. God has been man's explanation for everything that we did not know or understand, and the power behind all the forces of nature that we cannot control including ourselves and our ultimate fate...death. God is man's projection of the ultimate Human being, and the ultimate state of existence. God is as far as man can see or hope to grasp.

THE HUMAN GENOME IS FULL OF "GARBAGE"

The HUMAN Genome

Interestingly, our human genome is literally filled with "junk" DNA, which has nothing whatsoever to do with being human. Amazingly, this "Junk" DNA actually comprises more of human genome DNA than what is used to make us who we are... humans. For example, there are multiple copies of LINE1 DNA that make up about 14.6 percent of the whole human genome. This DNA could not even vaguely be considered to have anything to do with being human. Line1 DNA has been estimated to represent between a thousand to six thousand bases long, with about 100,000 copies in the human genome. The only **purpose** of this **Hugh** amount of LINE1 DNA is to **replicate itself** and "survive", and it is doing a good job of that since LINE1 DNA is thought to probably have existed in the human genome for thousands of years. There is probably more LINE1 DNA in the human genome than DNA used to make humans the "smartest or most intellectually advanced creature on Earth.

In addition, there are other large segments in the Human genome that are totally irrelevant to the intelligence or being human, and are only there to enhance their own existence. This has been a stunning revelation that was only realized recently when the DNA of the human genome was sequenced and studied in detail.

Another example of totally irrelevant garbage DNA in the Human is labeled ALV, which is estimated to represent DNA between 180 and 280 bases long and is thought to be present in a million copies in the human genome. ALV DNA represents about ten percent of the DNA of the Human Genome. In other words, if this data is correct, there is more Line1 and ALV DNA in the human genome than there is DNA that makes Humans Human. The DNA bases which make humans the most "intelligent and advanced creature on Earth" is only a small minority of the DNA in every cell that is in the Human body. We Humans and most other "higher "creatures are full of viruses and other "junk" DNA. The Line 1 and AVL DNA in every human cell are there to replicate themselves and have nothing to do with Human intelligence or human existence. Humans have been highly parasitized by 'garbage' DNA. There are many more viruses and junk in the Human genome and other "higher creatures" than there is DNA necessary to make such creatures what they are! **This has been a stunning revelation of the sequencing of the human genome.** One of the major future Advancements in the human genome will be to delete the garbage DNA and replace it with DNA sequences that will enhance human intelligence and existence. This will allow a potential major advancement in human development. Replacing this garbage with useful DNA will open up the possibility of incredible advancements in Human development.

GCOS: "Many are called, but few are chosen."
However, there is a major problem.

The superlative God—all-knowledgeable and all-powerful—is what mankind has always desired to become. We have eaten the fruit from the tree of knowledge of good and evil, and there is no turning back. Gods are what mankind aspires to become, and future Gods are what man will evolve into using intelligent directed evolution (IDE_H). **Man will be the remaker of mankind**...perhaps many times over to reach such a lofty goal. This is the future of God! This is the future of mankind! The journey will be painful and long, but at the end of the journey, mankind will be the beatific vision of ultimate knowledge and ultimate power. Perhaps in a geologic blink of an eye, man could evolve into the God of all the creatures on Earth including the evolved monkey that invented even the concept of Gods to begin with.

GCOS: "Can you see the promised land?"
Arthur C. Clarke, science fiction writer famous from the movie *2001: A Space Odyssey* has suggested that a sufficiently technically advanced society would appear as magic. Michael Shermer, author, and publisher of *Skeptic Magazine* has paraphrased and extended this concept and noted that a sufficiently technically advanced civilization would appear as Gods. That concept can be generalized even further, and realistically applied to both current and future mankind. A sufficiently intellectually advanced mankind would develop the knowledge base and technical capability that would at least approach many if not most of man's current concepts of Gods. Such sufficiently advanced intelligent beings would, in fact, be Gods (not simply appear as Gods). They would be real Gods, at least as God-like as all the Gods created by mankind in the past.

Will Homo sapiens ultimately evolve into the God of mankind's lust...of mankind's dreams?

Remember, we have repeatedly noted that to understand or discuss God, we must first define: "What is a God," otherwise any consideration of "what is a God" or of having God-like qualities or becoming Gods degenerates into babble! Here we are using the concept of advanced intelligence as the defining property of a God, with all the other "God characteristics" literally being subsumed under intelligence. The Clarke-Shermer conjecture of a "technically advanced" being (or civilization) is clearly integrated into the concept of advanced intelligence. We have already noted that even the understanding of any basic concepts of a God requires advanced intelligence. In fact, as we have repeatedly pointed out, mankind is the "God creator." All the Gods ever created on Earth were created (or invented) by mankind. In addition, mankind even created the initial basic concepts of a God. Mankind is the only living creature on Earth that interacts, worships, believes or even understands the concepts of a God. Language, writing, mathematics, and abstractions like God, Heaven, and the tooth fairy arise out of the mind and intelligence of mankind. Mankind lives in the world of reality like all other creatures on Earth, but we have also ventured into the realm of the metaphysical—the land of thoughts, abstractions, mathematics...the land of the Gods.

We have generalized the concept of God to include present day mankind as a pitiful, pathetic, inadequate example of a God, but a God, nonetheless. Mankind is a God, but we are clearly at the low end of the totem pole of Gods, with the lower limits of intelligence, knowledge and technical ability expected of any God. However, we have projected the future paradigm of **mankind's evolution to be at a critical point, in which the random, blind mutations with "natural selection" (survival of the "fittest") will no longer be the driving force of evolution. Instead, the new paradigm of evolution will be driven by Human intelligence**, which will increase geometrically, and be self-generating, and "natural selection" will be replaced by intelligent choices of future genes and future beings that will continually evolve toward advanced Gods.

The future evolution of Homo sapiens will asymptotically approach the all-knowing, all-powerful God of our forefathers, but these future Gods will be real, not the stone carvings on the columns or walls of Egyptian temples or mythical Gods—like Zeus of ancient Greece, or the Roman God Jupiter or the invisible God of Abraham, or the son (Jesus) of the invisible God of Abraham, and the list goes on and on.

God is what man has always desired to be, and Gods are what man is now and ultimately will continue to evolve into…in the future. However, for 21 st century man, our "Godship" only extends to perhaps C. elegans, bacteria, insects, worms and other "lower life forms," and although we are beginning to understand such creatures (and their genetic makeup), we clearly do not control them, and no one (at least no one in their right mind) is interested in insects praying to mankind or worshiping Humans as a God. Homo sapiens per se (in our current form and intelligence) is not the ultimate God or the Gods of our forefathers. However, even at our lowly level of intelligence, mankind did invent the basic concept of Gods. No other living creature on Earth has shown any evidence of understanding any of the concepts a God.

But Homo sapiens have had the dream…the dream of a God…all-knowing and all-powerful (even beyond the known laws of physics). Man has dreams of an afterlife…life without end. It is a magical dream, a powerful dream, and a dream that mankind is now trying to create into a reality. We have already significantly increased the average lifespan of mankind in the 20th century, but we are only recently beginning to understand the aging process per se as we have previously discussed (In a soon to be published book which is already written) **We are almost ready to "cross the line," and begin to modify our own genetics, and take over our own evolution. In fact, with the many changes that we have already made to our environment, we have already crossed the line!**

We are at a critical point in the evolution of mankind where we are acquiring the knowledge and technical skills to begin to direct mankind's future evolution. In the future, no longer will evolution result from blind random mutations, and "natural selection." **Evolution will be intelligently directed** by mankind, Considering the incredible amount of "garbage" in the human genome at the present time, we desperately need to take over the "evolution of the human genome". The incredible impact of this paradigm on the future of mankind will be stunning. We will eventually extend the lifespan (and health span) of mankind many folds perhaps, in time, even to the edge of immortality.

Yes, there are many naysayers who oppose any change in mankind, but instead prefer to pray to invisible Gods of their own making. Unfortunately, these individuals do not seem to grasp the fact that if their God created everything on Earth, and is in control of everything, then that God also gave mankind cancer, heart disease, strokes, and every disease and every "natural catastrophe" that occurs on the face of the Earth including aging and death. Mankind does not need such a God.

GCOS: "Thanks God."
Let us not blame a God for the realities of life on Earth and the plights of mankind. The reality is that life in general and mankind in particular has evolved through random mutations and all the other genetics and chemistry (some of which we have discussed), and God had nothing to do with any of it. However, we are now beginning to understand and even modify and control these processes, and that will soon give us the knowledge and tools necessary to control our own evolution and our own destiny in the future. There will undoubtedly be disasters in our future. Nothing is generally accomplished without paying a price. Going into space had its price in terms of the ultimate sacrifice...Human deaths. However, no Human ever born has not suffered the ultimate insult of death, brought to us by "Mother Nature" or if you prefer, the all-merciful, all-benevolent, all-good God of our forefathers.

GCOS: "Wake up mankind! You are living in a delusion. If you want an all-powerful, all-knowing God who will give mankind eternal life and a Heaven to live in., then you must create that God in reality, and not as some stone carving on an Egyptian obelisk or an invisible God in the Bible!"

Consider the concept that even at our current primitive state of Homo sapiens' evolution, our current intelligence, modest though it may be on an absolute scale of intelligence, is massive compared to almost all living creatures on Earth, such as the bacteria like Escherichia coli (which could arguably be assigned an intelligence of close to zero, since it has no neurons, and its "intelligence" is, in fact, at best evolved chemistry functioning to the level to promote its survival or consider the worm, C. elegans, with 302 neurons (which has a ballpark estimate [another name for a guess]) of an absolute intelligence of perhaps I_{ab} = 0.000003 (to arrive at this questionable estimate, I simply applied a comparison with Human IQ and the number of neurons in Humans versus C. elegans to give a ballpark estimate of C. elegans' "intelligence." It is clearly a very primitive comparison [estimate]. In fact, despite the presence of 302 neurons, some would argue that C. elegans has no intelligence. However, any comparisons of intelligence do not come easy.) There are many pitfalls in making any comparison. Some will consider this a totally inadequate estimate, and I would agree. However, what is clear is that C. elegans does not have an intelligence of zero like a rock or perhaps even a bacterium. According to current estimates, C. elegans is composed of 959 somatic cells, 302 neurons, 81 muscle cells, and 17,800 genes. Among the 302 neurons, 14 are neurosensory neurons that provide information about the environment (what in this case is essentially a form of smell). The IQ estimate offered is a "ballpark" estimate to be used only for primitive comparative purposes. If you want a different estimate or even zero, fine, that is your estimate and is perhaps as good as mine. We are trying to make some kind of comparison, which at the present time is difficult and inadequate. Feel free to make your own comparison. The simple fact is that there is a tremendous difference between the intelligence of C. elegans and Humans and a rock. (http://www.cals.ncsu.edu/course/zo402/celegans.html, http://www.pubmedcentral.nih.gov/articlerender.fcgi?artid=552904#ref8)

The difference in intellectual capacity between C. elegans and mankind ($^h I_{ab}$ = 0.000003 vs. 100 spans at least 7 to 8 orders of magnitude—perhaps ten to a hundred million). What would a creature (an entity) with an intelligence of one hundred million times (or even one million times) Human intelligence be like? Would such an entity qualify as a Clarke-Shermer God? No one can even begin to imagine what such an intelligence would really be like. Is there a limit to upper intelligence that cannot be surpassed. Such intelligence might even subsume the Universe, or it really might represent nothing more than some primitive intelligence in what is ultimately possible. **In fact, one interesting idea is perhaps the whole Universe itself is the ultimately an advanced intelligent entity that we have yet to be capable of understanding or communicating with. Perhaps the entire universe is a God. Who knows!**

We tried to give some "ideas or comparisons" of advanced intelligence in our discussion of intelligence, however, the effort was undoubtedly inadequate. For example, if we consider a future advanced Human (a God by some definitions) with a *Human* $^h I_{ab}$ of 1000 (roughly ten times the current average Human IQ), we anticipate that there would be a ten times improvement in memory. However, although such a projection for memory might be valid, it is simplistic, and it does not take into consideration all the "emergent phenomena" that has occurred in the development of mankind's intelligence such as the development of language. In other words, many of the concepts that we have considered as components of intelligence are, in fact, entirely new facets of intelligence that could not have been predicted.

Neither I, nor you, nor anyone else can predict the future facets of intelligence that may emerge. However, as we view the development of intelligence from jellyfish (with the first true neuron and sense organs (light-detecting "eyes") to C. elegans and up the intelligence ladder to man, it is probably safe to predict that there will be at least some if not many totally new ingredients in the intelligence stew that will rival language, writing, mathematics, the use of tools, or even the concept of God, etc.

However, the comparison of mankind's intelligence to C. elegans make man's current state of enlightened ignorance seem very God-like...at least in comparison to many simpler forms of life on Earth. In fact, IQ was never designed to be used as an intelligence marker for all creatures. Perhaps we should invent some other new scale of intelligence that could be applied universally. Mankind in many respects already fulfills Clarke and Shermer's criteria of a magical being that looks like a God in comparison to most of the creatures that currently inhabit the Earth. C. elegans uses its 14 neurosensory neurons to smell/taste dozens of environmental chemicals, which stimulates behavioral responses of avoidance and or attraction...is that intelligence or not (at least in its very primitive state). However, C. elegans likely has no real awareness of how many cells or genes compose its body or any concept of atoms or stars or any recognition of the general composition of the Universe, which is why Humans have raised to the level of a **Sentient Enlightened Metaphysical Being** (a SEMB entity or in plain language, a God).

If you are skeptical of mankind being a God or consider such a concept either ridiculous or blasphemy, then consider mankind to be a primitive or "first level" SEMB that can be evolving into an even more advanced SEMB or choose another term for mankind (like "an advanced monkey.") The difference between a God and a SEMB, or an "advanced monkey" is only a matter of definition or one of semantics. Within this context, a SEMB entity is just another word for a God, however a "real" God of sorts: simply, intelligence with specialized cells controlling the response of other (effector) cells, evolving to the point of advanced abstraction and awareness, where other emergent phenomenon again arises.

However, before we venture too far into the sin of hubris, we should also recognize the considerable limitations and folly resulting from our current level of intelligence. The list of Human egregious faux pas includes wars, killings, starvation in a world of plenty, environmental damage, outrageous social injustice, and every kind of indignity heaped upon our fellow man, many if not most animals and the environment itself, and the list goes on and on. Mankind and the Earth is clearly in desperate need for more intelligence!

When did we first achieve this "milestone" of advanced intelligent development? We will leave that to future philosophers and scientists to debate when they have the advantage of "hindsight," and they will be able to see the past much better than we can see the present.

What if we used genetic engineering and **IDE**$_H$ to increase Human intelligence by 8 logs …roughly as much an increase above current Human intelligence as current Human intelligence is above the worm, C. elegans; would that qualify mankind to be a "true God of our forefathers" …not just a "look-like God," but actually be a "real God"?

With mankind's current knowledge base and technical advancements, and with our achievements in language, mathematics, abstraction, and reasoning, it is relatively easy to fit even present-day mankind into the God (SEMB) category. In fact, it should not require much more of an increase in intelligence to qualify for the "God trick" than the current 7-8 log increase in IQ intelligence of Humans compared to the worm, C. elegans (using our rather crude comparison of "intelligence" based on the number of neurons). In fact, it is probably much higher considering the more recent realization that much of the human genome is junk, and has nothing to do with human intelligence.

The genome of man, E. coli and C. elegans have all recently been worked out. We now know every gene that constitutes these creatures, and…even the genes that constitute ourselves (I am not going to stop and defend this or enter the controversy that we are still trying to sort out which segments of DNA constitutes a gene or how much "garbage" is in our own DNA). We can grow E. coli and C. elegans in our laboratories. We have already genetically modified both creatures. In essence, we have recreated them as we please! We understand their place in the Universe, and the history of the molecules that make their very existence possible. Yet, the worm and the bacteria likely have no cognizance of the world around them or even any vague awareness of the Universe in which they live. In a very real sense, man could be considered a God compared to E. coli and C. elegans using some of the criteria that man has consistently used to describe Gods in the past, such as great knowledge, power, and understanding.

After all, mankind has at least some understanding of the world in which we live, and our place in that world, which far, far exceeds that of any other creature on Earth. However, ultimately, mankind's current "God-status" still depends on one's definition of a God. Since we invented the concept of God to begin with, we should also be entitled to manipulate it.

These comparisons and estimates of mankind's "God-status" does not spring from arrogance, but from fact. A statement that is true is not necessarily arrogance. Here we be...Gods...at least of a sort even now. In fact, one of the popular Gods of today was associated with a box called the Arc of the Covenant which was placed in the Hollie of Holies in the Temple in Jerusalem, but as previously noted the 10 commandments were in the Box, bur it is not clear where God was in relation to the

Perhaps mankind is not that far removed from the prophecy of Genesis 1:27, "So God created man in his own image," and that prophecy will finally be fulfilled. God and evolved man will finally be in the same image, Genesis 1:27 will be fulfilled, with only the small variation in the sequence and mechanism. We will be creating God (ourselves) in our own image.

We previously discussed some of the meaning and purpose of Human life. What is the purpose in life of the worm? What meaning has the life of the frog? The worm is our ancestor from about the time of the Cambrian period—about 540 million years ago. We share genes with the worm as with many of the other lowly creatures that creep and crawl upon the Earth. For Homo sapiens, perhaps the most important purpose and meaning of the life of the worm was to evolve to a higher level of neurological and functional development—to evolve eventually into Homo sapiens. How far is that concept of "purpose" from the Biblical concept that the purpose of creation was mankind?

Many individuals still have a considerable problem accepting the concept that man (or any life form) has "evolved spontaneously," and they insist on a God as a creator! Perhaps, the Universe itself is a creator God. The answer to that type of question is many orders of magnitude beyond my intelligence capability (and I opine, yours also). However, it is clear that the God of the Bible or of a man nailed to a cross in ancient Jerusalem is not that creator God nor are any of the Egyptian, ancient Greek, Roman or other currently proposed creator Gods of the past. These ancient Gods (and the concept of God itself) are the creation of mankind. The whole idea here is that there is a spectrum of Gods, and all these different concepts of Gods still fit under the umbrella of "Gods." They can be anything we want them to be! **Get used to it; we are Gods! In fact, we even created the concept of a God itself!**

Perhaps the real meaning and basic purpose of Human life is that we are merely the monkey that has evolved into intelligent creatures that are beginning to understand the Universe in which we live. From an anthropomorphic "purpose," the worm's "real purpose" was to evolve into man. Homo sapiens' "real purpose" will be to evolve into more advanced SEMBs (Gods). Has the whole Universe been "set-up" to evolve mankind? The answer to that idea is really a religious question, so I will not go there. However, we will note that the entire concept of God based on current and past religions is only a primitive prayer (projection) of most of the concepts being suggested here, put forward before there was really any understanding of the processes involved or mechanisms to develop such a reality.

In the 1950s there were constant sightings of Unidentified Flying Objects—spaceships. The question quickly arose, "Are there spacemen?" We now know the answer. **Yes, there are spacemen. We are the spacemen!**

Man has asked for thousands of years with our limited ʰI_ab, "Is there a God?" The answer is...**Yes! We are the Gods.** Primitive Gods currently, but with all the potentials of the Gods of our forefathers. However, we will achieve these advanced properties by understanding and advancing our intelligence and scientific capabilities, not by prayer or magic. Interestingly, as noted above, the human genome is full of 'garbage', but there appears to be enough DNA to control the garbage. (**Thank God or whoever is running this universe!**)

We were not created: "made the image of God"; instead, we (mankind) have created many Gods in the image of man. We have also created Frankenstein (chimeras) and Gods that are part animal and part man. We have even created Gods that are invisible. Almost every aspect that man has attributed to Gods are, in fact, Human traits or desires, which mankind has projected onto one God or another throughout recorded history. Even more egregious, we have imparted to these Gods enormous powers, and we then request (should I say demand) that they use these powers for our benefit. It is the absolute ultimate in Human arrogance and gullibility to propose that the creator of the Universe—a Universe so huge and so magnificent as to defy any real grasp by current Human intelligence, so complex that on a quantum level it defies common sense and understanding, and in reality disappears into probabilities—was created by an invisible God who formed a covenant (a contract) with a group of nomads in the desert, and had a son with a Human woman, and that son was nailed to a cross, and died for the Human sin of eating an apple (original sin). Yet, many balk at the concept of a future mankind with markedly increased intelligence being called a God!

Yet, if we use other criteria for a God such as understanding the Universe (rather than creating the Universe), development of considerable (even extreme) intelligence (compared to all other creatures on Earth), and the possession of a huge database of knowledge (which would translate into the power to manipulate our environment), and with even the potential for immortality or at least a significantly longer life expectancy than is the current model; then the descendants from a monkey might one day qualify…to at least approach the incredible magnificence of the God of our forefathers. After all, mankind is the one that determined (invented) all the properties of Gods to begin with.

Even the Bible notes that man is a God and was a God two and perhaps three thousand years ago. Psalms 82 (KJV) states: "I have said. Ye are Gods; and all of you are children of the highest." In fact, Jesus used this Old Testament statement when he was being pursued by a mob trying to stone him for blasphemy (because he had said that he was the son of God).

"Jesus answered them, is it not written in your law, I said, Ye are Gods?" (KJV, John 10:34) Jesus was arguing that it was not blasphemy for him to call himself the "son of God," since we are all Gods or the sons of God. Jesus cements the argument with: "**…and the scripture cannot be broken.**" (KJV, John 10:35) Apparently the mob was not "buying" Jesus' argument: *"Therefore they sought again to take him: but he escaped out of their hand."* (KJV, John 10:39).

The journey continues. I am again remembering the last few lines of Robert Frost's poem:

And miles to go before I sleep,
And miles to go before I sleep.

The concept of man evolving into a God may sound implausible. There are no weak and puny Gods. Gods are in the business of doing impossible things—miracles (another word for magic) … like saving lives! You are sick and dying. You pray to God…to save your life. You are dying of pneumonia. You are gasping for every breath. "Behold an Angel of the Lord appears," and gives you…a shot of penicillin, and the infection clears. Your life is saved! A miracle, no less!

GCOS: "Praised be to God."
"What about the penicillin?"

GCOS: "God provides."

"But what if someone called a doctor—no Angel, just a regular old doctor came by and gave you the same shot of penicillin, and you lived. Is that a miracle?"

GCOS: "God provides."

"Perhaps you are right—Gods provide...especially when that God has penicillin."

Gods are powerful. Gods can do impossible things...like save or extend lives. In this particular case, the "God" was a mold that secretes a chemical called penicillin. Of course, the mold does not do this to save the lives of Humans. You can pray to the mold all you want, and you are not going to get any penicillin. The mold secretes penicillin to protect itself and its food source from bacteria. You got the lifesaving "God" power of penicillin because a Scottish physician left a bacterial culture plate on his lab bench in 1928, when he went on a two-week vacation. When he returned, he noted that the culture plate was covered by a "lawn" of bacteria...except for a halo area around a mold that had accidentally contaminated the plate. Fleming recognized that the mold was producing something that inhibited the growth of bacteria. Unfortunately, Fleming was not much of a chemist, and it was not until 1939 that two Oxford scientists (who were chemists), Florey and Chain, isolated penicillin. Penicillin has now saved millions of lives—like a God...and generally, penicillin has been much more reliable at performing miracles than praying to the ancient Egyptian God Osiris (remember, the Egyptian God of the underworld...perhaps the first God of the afterlife).

But it is ridiculous to even consider a lowly mold as...a God, even if it secretes penicillin, and has saved millions of lives. However, the men like Fleming that used intelligence to recognize that the mold was secreting something that killed bacteria, and others like Florey and Chain who extracted the use of the penicillin, took a "God-like" step. Mankind can now save lives just like a God with the knowledge and power of penicillin. Does that make us a God? Well, perhaps mankind has got his foot in the God door. It is all a matter of how we define a God.

However, when that little infinitesimal of knowledge and power mankind possesses is multiplied by some very large number…then the reality will begin to be indistinguishable from the God of our forefathers. God (remember Clarke-Shermer's conjectures of a technically advanced civilization would appear to less advanced societies as Gods). Miracles like penicillin are already an everyday phenomenon in modern medicine. Of course, we are only considering one property of Gods—saving lives. In essence, the use of penicillin is one of the "Gods of the Foxhole."

Let's see; Jesus Christ brought Lazarus back from the dead, and then there was the Marks Gospel story of Jairus' 12-year-old daughter who was also "raised from the dead."

That is two lives (supposedly) saved by Jesus. Buddha has not been reported to have prevented anyone from dying, but he was in the business of achieving nirvana…not saving lives (in addition, although he is treated like a God, he never called himself or considered himself a God). The God of Abraham…he did not save any lives either. In fact, he supposedly killed every man, woman and child, and animal, and every living creature on Earth with a flood (except for Noah and his family, and the animals that made it onto the arc) …according to the Bible. Then there was the unfortunate guy who tried to catch the tabernacle as it was falling over—killed instantly on the spot. Let see… Miriam was turned into a salt pillar…she looked back. Then there were all those people in Sodom and Gomorrah, destroyed. Then there were all the people of Jericho that we discussed…killed, every man, woman and child and all animals and living creatures in Jericho. This was done under the direct command of the God of Abraham. The God of Abraham seems to have killed a lot of people…but not many lives saved. The God of Abraham does not have a good track record for saving lives. Everything considered…that lowly mold (penicillin) is looking more God-like than ever, at least when it is in the hands of mankind.

I generally prefer a moral to a parable.

What is the moral here?

GCOS: "If you are sick and dying from an infection...pray to the penicillin mold."

"No...that really doesn't quite "do it" as our moral."

GCOS: "How about...if you get sick with an infection like pneumonia, take penicillin."

"That might be smart. It might even save your life... but it still does not quite 'cut to the core' as our moral."

The principle (moral) that we want to "dig out of here" is: "If you want a God to solve your problems—like save your life—you had better become that God." Historically, most Gods have a poor history of saving Human lives.

Remember one of our themes, **Made by Man.**

GCOS: "Ridiculous...How does a Human become...a God?"
Julius Caesar was declared a God in 42 B.C. (he had been stabbed to death on March 15 of 42 B.C.). Of course, most of us do not consider Julius Caesar to be a "real" God. However, stating your objection in 42 B.C. could have resulted in a fatal disease—like being hacked to death with those short little Roman swords. If you were living in 42 B.C. in Rome, common sense would have told you that Julius Caesar was a God. If you did not "get it" then the best advice would have been to keep your mouth shut...and pray to Caesar. However, the road map we have prophesied here does not involve voting for a God or building pyramids or building churches; it involves building laboratories, and understanding the world in which we live, and **most important in understanding ourselves. Man, the re-maker of mankind... again the principle is: Made By Man!**

Do not expect some omnipotent mythical God to "save you." Save yourself—you are the only God that you will ever have (and bring the penicillin). Santa Claus brings toys to children. You are no longer a child—you must go get the "toys" yourself.

To paraphrase Martin Luther King's great speech; We have been to the mountain. **There is a promised land!** Most of us will not get there…but our children will! Yes! They will be the Gods…Your future children will be "real" Gods, perhaps more like the Gods of our forefathers. An advanced SEMB or perhaps a Silicon Deus, the future Gods of Earth…it does not matter. They will both be our children, and they may even be the same or some amalgam.

GCOS: Praise be to the Gods!

The woods are lovely, dark and deep,
But I have promises to keep,
And miles to go before I sleep,
And miles to go before I sleep.
(Robert Frost)

GCOS: "Is this the end?"
"No, it is just the beginning!"

ABOUT THE AUTHOR

Melvin G. Dodson grew up in Miami, Florida and from a young age he was very interested in religious studies. He even received an award for religious studies while attending a Catholic high school in Miami. However, when studying these concepts more in-depth in college at the University of Florida as an undergraduate, and looking at the data, Dodson began to realize that mankind invented God rather than God inventing mankind, which is one of the major concepts of this book.

After that, Dodson went to the University of Miami Medical School. After completing his residency at Jackson Memorial Hospital in Miami, Dr. Dodson served two years in the U.S. Army Medical Corps.

Dr. Dodson went on to receive another doctorate degree in microbiology and immunology. He then spent almost 20 years teaching academic medicine at four major medical institutions. He later moved back to Florida and went into private practice. Dr. Dodson has always had a broad and continuous interest in the background of human development, including our origin, as well as the essence of Human intelligence. He believes that mankind is currently taking over our own evolution and will soon change our genes to cure many diseases, extend the lifespan of man to hundreds of years, and markedly and prominently improve the intelligence of mankind with IQs into the thousands. The results of these changes for mankind and the world will be dramatic! These concepts drove Dr. Dodson to write this book.

It is his hope that by reading this book, people will be more hopeful of their bright futures and the futures of their children, more educated to better understand the changes that are occurring in our world, and more inspired to seek the answers to the great questions of mankind's origin, development, and future. As you read this book, you will learn that Dr. Dodson's answer to whether there is meaning and purpose to human life is that there is no meaning and purpose given from above or on high. However, mankind can assign ourselves deep meaning and purpose, and we should do so for the betterment of ourselves and our world. Writing this book was one important objective that Dr. Dodson has assigned himself, and hopes that you will find it enlightening, inspirational and most of all thought-provoking.

Godspeed!

www.ingramcontent.com/pod-product-compliance
Lightning Source LLC
Chambersburg PA
CBHW052138220526
45471CB00004B/1430